MW00835327

The Road Back to Nature

The Road Back to Nature

Regaining the Paradise Lost

Masanobu Fukuoka

Japan Publications, Inc.

Published by JAPAN PUBLICATIONS, INC., Tokyo and New York

Distributors:
UNITED STATES: *Kodansha International/USA, Ltd., through Harper &
Row, Publishers, Inc., 10 East 53rd Street, New York, New York 10022.*
SOUTH AMERICA: *Harper & Row, Publishers, Inc., International Depart-
ment.* CANADA: *Fitzhenry & Whiteside Ltd., 195 Allstate Parkway,
Markham, Ontario, L3R 4T8.* MEXICO AND CENTRAL AMERICA: *HARLA
S. A. de C. V., Apartado 30-546, Mexico 4, D. F.* BRITISH ISLES:
*International Book Distributors Ltd., 66 Wood Lane End, Hemel
Hempstead, Herts HP2 4RG.* EUROPEAN CONTINENT (except Germany):
*PBD Proost & Brandt Distribution bv, Strijkviertel 63, 3454 PK de Meern,
The Netherlands.* GERMANY: *PBV Proost & Brandt Verlagsauslieferung,
Herzstrasse 1, 5000 Köln, Germany.* AUSTRALIA AND NEW ZEALAND:
Bookwise International, 1 Jeanes Street, Beverley, South Australia 5007.
THE FAR EAST AND JAPAN: *Japan Publications Trading Co., Ltd., 1–2–1,
Sarugaku-cho, Chiyoda-ku, Tokyo 101.*

First edition: August 1987

LCCC No. 86–81325
ISBN 0–87040–673–6

Printed in Japan

Contents

8

Preface to English Edition

I have chosen to call this book *The Road Back to Nature*.
But because there may be some uncertainty as to precisely
what I mean by this, I would like to begin by clarifying the
notion of "returning to nature" that I have in mind. To me,
this represents the effort to reunify God, nature, and man—
which have been split apart by mankind. Unfortunately, man
cannot be successful in his attempts to return to nature unless
he knows what true nature and God are.

Some people say that nature ravaged by man is still nature,
that the wasted desert lands left in the wake of human civilization are nature nonetheless. Yet, however far he may wander through such fields and mountains and however long he
may live in a secluded glen from which issues a mountain
brook, man can only gaze upon nature's outer shell; he will
never have access to its true inner heart and soul.

The soul of nature is also the will of God who dwells in
nature. This is not something that scientists can find by dissecting nature. Indeed, scientists are incapable even of knowing that they are in no position to understand the soul of a
flower in the meadow.

In the belief that they are exploring the root of life, genet-
icists extract and synthesize the genes present in the cells of
living things. But nature's soul does not lie hidden within
DNA. Nor is this where God resides.

Recently, physicists claim that the mental attitude embodied
by the Eastern concept of nothingness (*Mu*) appears close to
resolution within the realm of the quantum theory, and I've
heard it said that astronauts floating in space are able to sense
God's presence while in a weightless state. But God and
nature lie beyond the grasp of the human intellect. No matter
how much he dissects nature or with his own mind denies

the existence of human intelligence and calls everything *"Mu,"* far from being able to observe the real state of nature and grasp the essence of God, man only moves further and further away from both nature and God.

The fragmented and diffusive development of knowledge which expands outward without aim or direction has brought human thought to the extremes of confusion, recklessly splitting apart God, nature, and man—originally one indivisible—and leaving only a legacy of incoherent chaos. Science has gone on a wild rampage, and the global disruption this has caused is only now becoming clear. Quite frankly, there no longer seems any hope that man will succeed in returning to nature and reuniting with God.

I, who received a revelation of God one moment fifty years ago, was so taken aback by the vision I saw that I failed to advance along the road I should have. Instead, I turned my back on God and tried to follow quietly the path of a solitary farmer. In time, I came to call the road I had passed over "the road back to nature" and, growing arrogant, professed to practice natural farming.

Although I claim to practice natural farming, all I have really done is to haphazardly conceive of a form closest to the image of nature and grow crops in accordance with that form. Of course, while I felt that the purpose of this should always be the revival of nature and the manifestation and concrete expression of the God that lies hidden deep within nature, there was little hope that I, having once ignored the will of God, should find a proper way.

Even so, I am grateful that when spring arrives at my free-spirited farm, for a space of several days the cherry, plum, peach, and pear trees in my orchard, and the semi-wild vegetables growing beneath them, all break out into bloom, mixing with the green of the foliage. As I look out upon my orchard now, it is truly a sight to behold. The flowers of nature bloom of their own accord and scatter without care or concern.

Looking at the plum blossoms radiant in the morning light and the whiteness and blueness of the *daikon* (Japanese radish) flowers, and noting the beauty of the iridescent shower of falling petals, visitors to my orchard call this an Eden, a paradise on Earth. But as soon as they finish clicking the shutters on their cameras, they hurry off home to the towns and cities. Even though they call this place beautiful, people today no longer have the time or ease of mind to immerse themselves in such sensibility. Rather than indulging directly in the raw, unrefined beauty of nature, they return home with their rolls of exposed film and are content to enjoy the natural beauty captured in their photographs. They don't call upon their innate sense of beauty, finding instead a greater attraction in nature as an object for self-expression. Nature remains only as a means for cleansing ignoble man.

People bring home flowers and display them in vases, vying with each other in the art of flower arrangement. Caught up as he is with the image of himself represented therein, civilized man today is no longer able to see the flowers (God) in the fields.

Modern man, who drops the flowers in the fields to embrace the wildly and falsely blooming flowers of vast civilizations, no longer understands what it means to return to nature. Where is the key that will open up the road back to nature?

Man has persistently believed that by amassing knowledge, sharpening his judgment, and deepening his ability to reason, he is able to exploit natural resources, advance human culture, and bring happiness to himself. But intelligence and reason were nothing more than perverse pranks. The greatest enemies responsible for man's loss of his native aesthetic sense and of the understanding inherent to man (transcendentally perceived knowledge that is the basis of good sense) are the human intellect and what we call *reason*.

Reason and understanding are mutually antagonistic. They play opposing roles. The intellect attempts to open up nature, but succeeds only in closing it down because human knowledge is in fact nothing more than a cumulation of judgment by the human intellect. At first, reason appeared capable of becoming the means necessary for conversing with God, but instead it turned out to be a dangerous weapon that strips man of wisdom and brutalizes God.

The flower perceived innocently is itself divine nature, but when examined with the intellect, this is transformed into the cold flower of reason, the heart of which nature shuts out.

What I want to say is that God did not create heaven, earth, and the cosmos. Rather, when the Earth was born and the meadow flowers bloomed, the butterflies fluttered about, and the birds sang, God came of his own choosing to dwell there.

Instead of praying to God as a mighty power that reigns over the heavens, man should have frolicked innocently with this wonderful sprite, this angel inhabiting the fields. That was the shortest road back to nature and at once the Great Way back to the side of God.

> Look how beautiful, the flowers of the earth!
> This is the land where live the gods;
> A perfect, faultless, natural paradise.
>
> Now in the deep slumber of spring in my Eden,
> I dream a private dream of returning to nature.
> Here there is nothing that must be done;
> No effort is required, not even courage.
>
> But no one even bothers to look back.
> Will the road to nature fade again
> Into the mists?

April 1986

Preface to Japanese Edition

This book is an attempt to paint a true picture of God, nature, and man. Such an ambitious venture is beyond the powers of a slow-witted farmer. Yet, well aware as I am of this, there is a reason why I have chosen to write this juvenile book. *One day, while still a young man, I saw suddenly the totality of God.*

I have never revealed this before. Why do I here spit out these words which for almost fifty years I have kept hidden in the depths of my heart? There is more to my reluctance than mere hesitation. No, I have ordinarily sought at all costs to avoid uttering the word *God*. This is because I knew that man is incapable of speaking about God, or of understanding or believing in God. Yet now, I deliberately break this personal taboo as, with intense remorse, I dare to say "God" and await divine judgment.

Of course, on that day in my far-off youth when I knew God—from that day forth, I should have followed His divine will. I should have walked the proper road for man that had been pointed out to me. But at the time, I was just a stupid, good-for-nothing youth hopelessly corrupted by the secular world. Overwhelmed with awe and amazement at the indescribably glorious sight of God, I shirked my duty. Whether it was cowardice or arrogance I do not know, but taking advantage of the fact that God gives man no instructions, I turned my back upon Him and began to walk the road back to my own egoistic self.

Proud as I was then, I thought that I did not need to know anything, that I needed nothing at all. There was absolutely nothing that I had to acquire through effort. I had not the slightest qualms in declaring openly that time and space had nothing whatever to do with me.

Even as I saw people racking their brains and laboring away at the production of goods in order to secure a bit of space and time, I thought, "Let those who wish to advance do so. As for me, I have nowhere to search for. All that remains is to turn back and slowly make my way home. No, even the road back home exists no more. All I should do is live and enjoy this day."

I knew very well that at heart I was a fool. Well, let a fool live as a fool will. All I had to do was to protect this folly of mine. On the pretext that it suited me better to labor soiled with the mud of the corrupt world than to cruise through the proud and lonely heavens, I entered of my own choosing a life of escapism. I turned my back on both God and myself.

It was only fitting then that I was forsaken once again by God and treated by society as a lone stray bird. This, in spite of the fact that I had forseen it all from the start.

What happened, very simply, was that, taken aback by the vision of an ideal paradise before me, I turned around and bolted from reality. Although I had noticed that our world was a garden of Eden, I turned my back on this knowledge and, without fully realizing it, chose instead the road of a dilettante farmer tilling a lost paradise.

As I reflect back on it now, I realize that at the time I knew both the reality of life and the essence of death. Which is to say that I knew that one needs neither to labor in order to live nor to fear death. This cleared away all my worries and it became my intention to live a solitary life of serenity in accordance with the principle that it is enough to be alive. There is no denying that I spent those days in the greatest of bliss, enjoying each day with a child's delight.

But as time went by, being basically dull-witted as I am, I lost all that I had gained and tumbled back to earth, becoming once again a wretched fool. When I saw that I had lost my original naiveté and inspiration, I even begged to be readmitted to God's presence. But this was no longer allowed of me.

To others, I may appear as someone who has lived serenely

with nature and has carved out his own path while remaining dedicated to the way of natural farming. In my heart I too secretly desired this, but the truth has been just the reverse. Impatient with my own daily imbecility, I yearned after a God I was unable to forget no matter how hard I tried. Each day, I struggled with people and showed myself off to the world. I loved people but was unable to love them enough. Even as I lamented the world, I was at a loss as to what to do. I lived an idle life. Those were days of constant anguish and discontent during which I blamed myself and reproached others.

Tossed about by the contradictions within and without, I was a disagreeable presence even to those in my family. Come to think of it, over the course of these decades, I have not lived even a single day at peace. Although this is fitting punishment for one who turns his back on God and betrays himself, today I am tormented by heartrending grief.

The reason I lay bare here everything about my past is that I dearly wish to atone for my errors and make the fullest possible use of the few remaining years I have left on this earth. At the same time I want also to search for a way to preserve the light of natural farming that has managed somehow to remain burning to this day. This is because I have come to sense the danger that natural farming may vanish as just one transient farming technique.

I wish to make it clear that natural farming is not limited to my own humble experiences; it is not something that an individual can establish or that can be perfected by man. I sensed the will of God, but this task was beyond my powers.

Natural farming is nothing less than a true way of human life that revealed itself suddenly to me when I learned that nature is always a total perfectionist, that it is the image of a God who faithfully practices absolute truth, and that man cannot survive away from nature's embrace. Of course, the actual methods and general principles were revealed to me during that brief instant.

Natural farming must always be started and developed

according to God's instructions. All man ever had to do was to listen humbly to the word of God and lend just a little bit of a hand. Unfortunately, I was off in a trance and, caught unawares, failed to hear the proper response. I was not able to become a faithful servant of God.

Natural farming always appears to be incomplete, but it is at all times perfect and complete at the place of God. Acutely aware that, powerless as I and others are to do any better than merely imitate, I may have neglected the work of establishing natural farming because I did not call upon others and made no effort to create followers and students. However, I never once lost confidence in the fact that natural farming was possible.

God in all ages appears as a discontinuous continuum. Natural farming too, since antiquity, may have arisen and vanished, and risen again to flourish.

Natural farming is one of the spiritual lights that must be kept burning throughout the night. In this age in which we live, it is possible that if this light dies out, it may never burn again.

Natural farming is more than just a way of farming. It is without question the one and only path that remains for humanity to continue to survive on this earth. It must not be allowed to come to an end.

Whatever becomes of this old farmer is of no concern.

Even if the going is steep, we must lay out the route by which our children and grandchildren can continue to live on this beautiful planet of ours.

God has left man to his own devices; he has abandoned man. If man does not save himself, no one will do this for him.

June 1984

Introduction:
Man Doesn't Live by Bread Alone

Here are some excerpts from an interview I had at Shunjū-sha, publisher of the Japanese edition of this book.

*   *   *

Q: For the last twenty or thirty years, we've been hearing how rapid progress in science offers hope for the future. But more recently, with the famine and starvation we've witnessed in parts of the world and the rising fear of nuclear war, concern has grown that the days of humanity are numbered. What do you make of all this, Mr. Fukuoka?

A: Well, we have to ask ourselves what it is that man lives by. Christ said that man doesn't live by bread alone. I think that those words hold the key to whether mankind has a future or not.

Q: Is it being too optimistic to believe that man can survive if only he's able to maintain sources of food and energy?

A: Man has no idea what it is he lives by, and he's got no idea what to eat and what to depend on to live.

Q: Do you mean to say that simply knowing scientifically how man lives is not enough?

A: Does he even know this scientifically? I think it is fair to say that science doesn't have the vaguest idea what real food is. Science can't tell us why, how, and in what way man eats and should eat. It can't tell us the basis and cause for being alive, and it is useless for explaining or even describing the true source, meaning, and goal of life.

Are you yourself confident that you are correct in your own way of living, in your yen for life? Do you know for certain why you must go on living?

Q: Of course, I don't believe that all facets of life have been revealed to us, but isn't it fair to say that no other animal knows as much or lives as surely as man does?

A: Following that same line of reasoning, the foxes, the badgers, and the birds haven't the faintest idea what to eat or grow; to man, theirs is a very fragile, marginal existence. Well actually, it's just the other way around, isn't it? These animals are the ones with the most reliable wisdom and the most dependable way of life.

Q: In what sense?

A: They live each day fully without the least hesitation or uncertainty.

Q: I see. But man is no insect or worm. Even a life lived with uncertainty and anguish is okay. Why, man derives a certain pride from his passion for a human way of life. And then there is the view that man jumps at the chance to engage in the struggle to prolong life in the hope of assuring the eternal prosperity of humanity, something which other forms of life are incapable of.

A: People who think that way are convinced that the care-free, three-day life of the mayfly is meaningless; that even though a human life may alternate between joy and sorrow, this has more significance. But that's the same as saying that a stormy inferno is more interesting than a calm, peaceful paradise. To begin with, one can't measure the importance of life in terms of its length.

Q: But that very question of the length of life is most important. Ever since he emerged on the face of the earth, man has sought earnestly, has begged and supplicated, for eternal life and eternal youth. Isn't it conceivable that this desire stimulated the cerebrum and became the driving force behind the advance of the human race?

A: This long and pathetic desire for eternal life and youth was a longing for tragedy. Before fearing death and praying for life, man should have determined why it is that he came to fear death in the first place.

The sparrow does not fear death. But people, as they live on day after day, are always confronted with the specter of death. Why, and starting when, does human life slide into an existence in dread of death?

At first, human life neither issued from nor was constrained by death. Death is basically unrelated to life. The rice plant withers and dies each year, but the grains of rice go on living. Life is handed down from one generation to the next, and constantly born anew. Today's life ends today. The me of today dies today. I am not the same tomorrow as today. Today's life should be put into order today. To be alive means to live this day to the fullest. That is the only way to live.

Q: Are you referring to something like a "discontinuous continuum"?

A: I suppose that is one way of putting it. Life is both a discontinuous continuum and a continuous discontinuum. One must abandon oneself each day. Even though every day is a new beginning and is followed by a tomorrow, there is no tomorrow for this "me."

Q: Doesn't that make life empty and meaningless?

A: On the contrary. Don't you think it more barren an existence to regret each day one's death on the morrow and live with lingering attachments, praying and hoping day after day that there will be a tomorrow?

Q: But can't one subscribe to the view that there are two sides to this world of ours—a bright and a dark side—and that some live a light, carefree existence on the surface while others sink to and inhabit deeper layers?

A: What you're talking about is people being entranced by two false roads. Unable to select one way or the other, they dangle in between, leading indeterminate lives. People fail to notice in the words "man doesn't live by bread alone" Christ's rebuke that it is not with bread that we live. Those unable to dedicate themselves to the proposition that "I can live without bread this day" are unable to live even with bread. Without bread, they immediately begin to scream and howl as if they were about to die.

Q: Do you mean to say that this very day is a tragic day on which we must be prepared to abandon ourselves?

A: That tragic resolution of which you speak is already a bundle of egoistic self-attachment. It would be better to take an afternoon nap.

Q: But if one takes a nap, there's no chance for spiritual progress and I suspect that the only result will be that one strays down the wrong path. To tell the truth, the more I listen to what you say, the less I understand. . . .

A: Isn't it enough to realize that the more you think the less you understand?

Q: What do you mean?

A: If you stop thinking, you'll understand.

Q: Aha! I think I see what you're saying now.

A: You should know that there is a world of difference between thinking you understand and actually understanding.

Q: Well, for example, it's possible that only those starving children in famine-hit African countries can appreciate the difficulty of eating a piece of bread. I imagine that those of us in modern societies overflowing with goods have forgotten the origins of the food we eat. Perhaps one way to resolve this would be to take a look back at the primitive ages and examine what it was that man first started to eat.

A: That may appear to be a scientifically proper approach, but it cannot show us what it is that man lives by. Nature is in constant flux. The past is gone and over, and such an approach does not apply to the unknown future.

Q: What then should we use to guide us?

All One Has to Do is to Live: As I noted in the Preface, when I was still young, my eyes were opened for a brief moment; you could say that I became aware that I was alive. I realized that all I had to do was to live. For the first time I saw how wonderful a thing it is simply to be alive.

$$*\qquad*\qquad*$$

Billions of years passed after the dawn of creation. Bacteria arose on the earth's surface, then vegetation flourished, and animals and man emerged. All arose and developed naturally. All things in the universe undergo constant change together;

there are no exceptions. Believing himself to be the crowning achievement of God—made in God's own image, mankind set about to fashion the future with his own hands. But millions of years ago, the monkey who, by such reasoning, was the most advanced and highly evolved of the organisms on earth never claimed to be the lord of creation or made in the image of God. It simply lived in nature and didn't worry a bit. Only man today wonders what he should eat and how to live; only he frets over tomorrow. We should be concerned first with why, living as we do in nature, we are so reluctant to entrust our future to God.

Clever man does not know life; he agonizes over death and asks, "What should I eat?" I generally do not know what word to use when I talk of human life, just as I have difficulty expressing the notion of death. Whenever I try to say what one must live by, or, stepping back a little, when I try to explain in concrete terms what food to live on, I always fear being understood in a narrow, limited sense and forced to give an explanation that can only be specious.

This is why the only explanation possible is to say that all one has to do is to live. Pressed further, all I can do is explain as well as possible what manner of living is true to man by resorting to some vague, abstract analogy: "The sparrows live by picking at the seeds in the grass growing over there. All man ever had to do himself was simply to live." That is about the only explanation I can give.

People stumble right from the start by asking themselves what they should be eating. They turn inquiring eyes to the form, quality, and value of food, wondering what they should reach for first and what is important. Gradually, a few specific foods are selected as necessary for man from the large number present in nature, and the rest are cast aside. This thinking grows even worse, culminating eventually in the conviction that there is no harm in growing and eating whatever one likes.

First there is nature and there is food, and in the midst of this lives man. That was the original state of the world. But

the moment that people hearken to the view that first there
is man and he produces the crops of his choosing, humanity
is transformed into an arrogant lord who commands nature.

Man believes that he has spared and made wise use of
nature. He thinks that human intelligence is superior to the
wisdom of nature and God. But although humanity can
learn from nature, it is not able to control or guide nature.
One could say that nature bears the wisdom of God.

It has taken nature five billion years to create plants,
animals, and man. How can scientists hope to know what
organisms to create to supercede man? Man shouldered the
hardships of growing crops from the moment that he began
to think he could grow food for himself with human knowledge. He has become an animal that can survive only by
processing and cooking the food that he eats. It is like spit-
ting up at the heavens. These efforts have had the effect of
destroying nature and ravaging mankind.

Of course, I have no intention of saying simply to stop
processing and cooking food. It is just that I am worried
over the process whereby precious food is transformed by
false human knowledge into evil food, degrading and corrupt-
ing humanity. Human knowledge has passed beyond the
bounds of naturally derived knowledge. I reject human knowledge that deviates from the wisdom of God. I am afraid that
mankind may refuse to live in an inhabitable nature and may
cut himself off from the future with his arrogant intellect.

Creation is like a magnificent orchestra playing the sym-
phony of nature. Humanity should have been content as one
member of that orchestra.

Bored with just watching the natural drama about him,
mankind has been drawn toward a stage where he can play
a one-man show. Well, by tracing the effects that his disregard
for nature and the progress made in the foods that he promoted have had on his own destiny, we get a good idea of
where exactly this absurd struggle leads.

Wild, primitive wheat arose on the Mesopotamian plains in

the Middle East. Wild rice is said to have originated in southern China, Burma, and remote parts of Assam. Rice also reportedly existed since antiquity in the Saharan region of Africa. Ancient man who settled in these three great birthplaces of grain began cultivating the wild rice and wheat that grew there, in this way coming by an abundance of food. It was here that the Mesopotamian, Chinese, and Egyptian civilizations were built up. But today, each of these regions has been totally transformed to desert. All that remains are vast, desolate ruins. Why should this be?

I strongly doubt that this was the result of changes in climate and desertification that created food shortages, scattered people and destroyed civilizations. What actually happened was that trees were cut down in the name of civilization. Advances were made in farming methods and slash-and-burn agriculture adopted. All this depleted the soil, causing a decline in vegetation and setting the process of desertification into motion.

Up until about twenty years ago, I was optimistic. Even as recently as ten years ago, I thought that if our supply of energy runs out, mankind will be able to get by if only he is willing to practice natural farming. But lately, I realize all too well just how wishful such hopes have been.

Human intelligence has overdeveloped and become sidetracked. Today, man has metamorphosed into a creature that can remain unconcerned even if it loses sight of and destroys nature and God. It is fair to say that man's excessive confidence in his own intelligence has caused him to lose sight of, or rather to destroy, true nature and true food, true life, true God, even the true image of man himself.

With nature ruined, true nature is nowhere to be found anymore. Food worthy of being called the staff of life can no longer be had. There just are no natural children around any longer who live according to their instincts. It looks to me as if natural people, natural farming, and natural diet are all receding and vanishing at an accelerating speed.

1

America—
Land of Plenty?

Why has California Turned to Desert?

Up Above the Clouds ──────────────────────

In July and August 1979, a Japanese farmer who had never
left the borders of his own country paid America a visit.
Although he went there with no particular business in mind,
it turned out to be an extremely fascinating trip.

I had never flown in an airplane before. I can't say whether
it has anything to do with my having flown at 30,000 feet,
but my view of things has broadened quite a bit and so today
I may talk of big and important things.

I boarded the plane expecting it to feel something like
Songokū, the legendary Chinese monkey-sorcerer, flying above
the clouds. I thought the experience would be pleasant and
even exhilarating. What I found out, however, was that the
sight above the clouds is indeed magnificent, but when you
look out the window, there is nothing below. You feel as if
this hunk of steel is just floating up there in the air. There
is no sensation at all of flying. The only clue you have is that
the clouds over yonder keep coming this way.

If that were all, it would be okay I guess, but after a while,
it was time to close the windows and run the in-flight movie—
some gangster flic. It felt as if everyday life had been brought
on board the plane and we were watching a movie in our
own living rooms. Everyone sat in silence with a look of
boredom on his face.

I discreetly raised the shade on my window a crack and
looked out. Since the view is from so high, I would have

imagined the feeling to be a splendid one indeed, but I didn't feel that way in the least. Instead, I felt like a frog plucked from a pond, or a sheep, perhaps, that has been boxed up in a crate and is being carried off somewhere.

The invention of a plane that flies over the Pacific for nine full hours like this leaves one with a keen sense of conquest by science over nature. But a doubt immediately arose: "Has man indeed conquered nature?" I got the feeling that nature is looking on with indifference from some unrelated place. If there is such a thing as God, then it seemed to me that 30,000 feet into the blue is where the showdown between nature, man, and God takes place. The idea gave me a thrill and kept me thinking all the way to San Francisco.

When the plane reached San Francisco air space, the first thing that raised doubts in me was the vista of single trees standing alone on the bleak, yellow land. This contrasted sharply with the mixture of trees and rich vegetation we see growing in Japan. It was to me a very strange sight. Looking out the car window as we drove from the airport to the city of Berkeley, I noticed that the mountains along the way were also the same drab, sandy color. The topsoil had washed out, leaving bare rock exposed.

Upon asking about this, I was told that these mountains had at one time been mined for manganese, which is why they looked as they did. That hardly seemed a convincing explanation, though. I stopped at Berkeley that evening and, starting the next day, was taken on a walking tour of the University of California campus and other local sights.

The Californian plain is an endless stretch of dry, yellow land. Hour after hour, no matter how far you drive, that is all there is. The scenery dosen't change a bit. I found it odd that the grasses covering the plain were amber yellow rather than green.

The land is extremely poor and covered mainly with forage grasses such as foxtail and wild barley. Isolated trees of

several types such as one might expect to find in a desert
are scattered over the yellow plains. From time to time, a
large field of tomatoes or some other crop measuring hundreds of acres in size would emerge suddenly. But these were
always irrigated. Wherever there was green, water was invariably being drawn for irrigation. Without irrigation, the
fields turn into parched grassland.

As for the cattle, I would have thought that they would be
allowed to play leisurely in a rolling green pasture. But that's
not the case. The green grasses are grown in one special field
and the cattle left to graze wearily in the middle of a scorching
field of yellow grasses.

The city of Berkeley and the UCLA campus, on the other
hand, are carpeted with a solid layer of green. The city seems
a very beautiful place, but that beauty is the artificial green
of lawns and protected trees; it is not the green of nature.

I had my doubts that I saw was the real Californian nature,
so I spent a day touring paleobotanical gardens and natural
parks to look at plants that had inhabited California long ago.
Most of the forty days I was in the United States were spent
essentially looking at the grasses at my feet.

Traveling through the mountains on the outskirts of San
Francisco, we sometimes came upon large groves of eucalyptus trees. All the large trees were eucalyptus. But this is
not a tree native to California. It comes from Australia.
The reason this is growing so hardily here is that there are no
native American trees around to speak of.

I doubt that the crytomeria and cypress trees on the cam-
puses originally grew there either. As soon as one leaves
town, the land becomes bleak and yellow again. San Francisco, Berkeley, and Los Angeles are like artificial islands in
the middle of a desert.

I was surprised to find several plants native to Japan
among the grasses in this parched land and began to wonder
what this meant.

The next day, I was taken to visit a place called the Zen Center on the San Francisco coast. Started by the Zen teacher Shunryū Suzuki Roshi from Japan, this was later continued by Americans. When I visited it the Center had 400 members; forty priests—both men and women—lived there. They meditated morning and evening, and during the day tended a small, half-acre vegetable garden in which they grew the food they ate.

I doubt that very many Zen temples in Japan still grow their own food. But in America, I was told, there are dozens of Zen centers such as this one. The 400 members consisted of working people, students, and others who come and commute to work while undergoing training here. Some come to stay for one or more nights, camping and working in the fields. Their pursuit of thought and enlightenment is intricately combined with the farming life. I watched with great fascination.

The people at this center practice a form of organic farming, but concentrate largely on spices, growing only a very limited selection of vegetables. The garden is located at the bottom of a valley encircled by eucalyptus trees. Yellow, bare-topped mountains surround the site on all sides. The land is desolate and almost entirely given over to foxtail. A little bit of greenery can be seen, but these are all small bushes no more than five or six feet in height and looking more like desert vegetation than anything else. There are no useful trees about.

I was asked if rice can be grown on the land there and whether the method being used to raise vegetables is okay or not. When I took a look at their gardening tools, I found that these are built for and rely on the physical strength of Americans. The spades, hoes, and other tools were all inefficient. So I tried showing them how to properly use tools such as the hoe and scythe. But what really struck me was the small number of vegetable types they grew.

Curious as to whether these blighted yellow mountains were California's true nature, I examined the grass by the side of the road on our way to the coast. In the midst of this yellow grass I found a plant that surely was related to the wild ancestor of the *daikon* radish, as well as weeds common to Japan.

Then, continuing on down to the shore, I saw an area on the mountain to my right covered with a green forest. Fifty years earlier, trees similar to Japanese pines were planted there. Today, this is the site of a luxury housing development. The mountain on my left, however, was a barren desert. Here, under identical climatic conditions, was an expanse of green on one side and a desert on the other. Why?

The conclusion I reached was that California was not originally desert at all. Something had probably caused it to dry out. If so, then it just might be possible to revive this land.

The Spaniards Brought Bad Grasses ———————

We visited a redwood forest about twenty minutes inland from the coast. This was a virgin forest with trees two or three hundred years old standing close together, many of them measuring twenty or twenty-five feet in circumference. During the last ice age, a few spots in California were spared from the glaciers which wiped out the surrounding area. These became the "glacier forests," where gargantuan trees, some two thousand years old and measuring 400 feet high, stand today.

There was a tribal Indian chief here who looked about eighty years old. "Are you the guardian deity of this forest?" I asked.

"That's right. Say, you just said something very nice there," he answered with a happy grin. He spent a great deal of time showing me around. I learned many things that day. (On my return to Japan, I received from him a handmade mug made from the top of a three-hundred-year-old redwood.)

I asked whether this area had always been forested like this and he replied that it had. The forest was preserved as it had been two hundred years ago and made into a national park. A narrow road passes through the park, with a rope strung along either side. There are no other facilities, not even a single bench.

Outside the forest, just ten minutes away by car, the land is parched desert. But here, we have the complete reverse—a thick, lush forest. About one-third of the undergrowth consists of vegetation similar to that found in Japan. Just think of it. Here, in the middle of an American desert we have a sacred forest where Japanese grasses grow among the underbrush. A little hard to believe, isn't it?

Since I had been told that this forest was pristine and untouched, I was curious. "What was California like long ago? Something must have gone wrong sometime." He said that he thought things had gotten screwed up when the Spaniards arrived and began raising cattle.

After investigating for myself and later talking with some experts, I came to the personal conclusion that the pasture grasses the Spaniards had brought over with them contained foxtail seeds, and it was this that dominates the vegetation throughout California. The reason foxtail dominates other vegetation is that the seeds set and mature around June. Normally, after one plant reaches maturity and withers, other plants grow up in an ongoing seasonal succession, but foxtail grows to such a great density that other plants are unable to take well in the same ground. That is why the fields and hills in California have all turned yellow.

The seeds have irritating bristles. When these catch on clothing, they cannot be removed. I was told that dogs and cats sometimes have to be operated upon to remove the seeds, which dig into the flesh. These seeds were spread by birds and beasts, turning the entire region into a parched, yellow grassland. If the temperature is thirty degrees Centigrade (86° Fahrenheit), the heat reflected from this ground surface raises it to forty degrees (104°). Such a rise in temperature turns the place into a scorching desert.

From this I surmised that the vegetation in California underwent a change when the Spaniards brought over new grasses with them. I had the feeling that these later had chased out the existing mix of grasses, and that this had altered the temperature of the.region and triggered desertification.

I was still turning this idea over in my mind several days later when I was invited by the head of the state environmental agency to give a talk before thirty officials at the state capitol in Sacramento. At the capitol, I was ushered into the director's office, where I found a tall, slim young woman who was the number two person at the agency. I spoke with her for a full thirty minutes before the meeting began.

As I was sitting down, she cleared away a rock sitting on top of her desk. "That's a strange stone if I ever saw one," I thought.

"Is that a Californian rock?"

"No," she said, laughing loudly. "This is from Russia."

"Well, well," I said with a chuckle. After a short pause, I continued, "Since arriving in California, I've seen a lot I don't understand. What I mean is, although the land is a virtual desert, there are plants here similar to the grasses we have growing in Japan. What type of parent rock is there in California?"

"You know," she replied, "I was originally a mineralogist." Bringing over a large, heavy book to show me, she explained that the parent rock for Japan is the same as that for the region around San Francisco. Moreover, Hokkaido and the nearby islands have the same parent rock as southern Canada. The Same is true for Siberia and Alaska, and for Southeast Asia and Mexico. The distribution of parent rock is exactly the same on both sides of the Pacific. She also said that some specialists believe that the Pacific had once been a continent, and that when the mountains erupted, the lava flowed east and west, creating two separate land masses.

Japan has Mt. Fuji and California has Mt. Shasta. Both are large volcanos and similarly located. This, plus the fact that the grasses and parent rock for each are the same could very well mean that long ago Japan and California were part of one land mass.

The greatest difference is that Japan today has four seasons. California, on the other hand, has just winter and summer. There is no spring or fall, and very little or no rainfall. If the parent rock and grasses are identical, then one would expect both areas to have a similar climate, including similar rainfall. But at some point, California became a desert and Japan became a temperate climate with four seasons.

This is what we talked of before the meeting. As a result, my conviction deepened that nature as it exists today in California is not true nature at all, but rather a land and climate that was probably altered at some time by men and machines.

The Rain Falls from Below ─────────────

I jumped right into this topic at my talk afterward.

"I've kept my eyes glued on the passing scenery all the way here from San Francisco," I told the officials before me. "As soon as you leave San Francisco, the color of the land turns straw-yellow. The transition to a desert climate is very clear. Then, as you drive into Sacramento, you find the place totally covered with green trees. The entire city is a garden, with flowers, cactuses, and trees planted everywhere. Why, it's just like an oasis in the middle of the desert. Sacramento is a beautiful city, but there is something artificial and unnatural about the vegetation here."

When I asked whether Sacramento had always been like
this, my question elicited quite a response. One person told
me that proof that it may not have been can be seen in some
very old houses still standing in the city. I was taken later to
visit one of these. A stairway entered directly onto the second
floor for direct access to the interior when the floodwaters
didn't recede. This house remained as evidence that two or
three hundred years ago, the town of Sacramento, which today
lies in the middle of a desert, saw some mighty big floods.

We often hear that scarce rainfall is typical of a continental
climate. Perhaps rain falls from above meteorologically, but
philosophically it falls from below. If there is vegetation on
the ground, then water vapor rises here, condenses into
clouds, and falls back down to earth as rain.

Deficit Farming

This land was transformed into endless yellow stretches of
foxtail grasses. Clouds no longer form here, and rain no longer
wets the ground. In such an environment, modern agriculture
grew increasingly mechanized, until farming methods were
developed that rely heavily on chemical fertilizers and pesticides.

Walking about on the land and digging for myself in the
soil, I came to the conclusion that the soil in California had
not always been poor. Even now, the land is not lean, but
the soil near the surface is very seriously depleted. This is
the result of irrigating the fields and working the soil with
twenty- or thirty-ton machines four, maybe five times a year
until it becomes hard as clay. The sun bakes and dries this,
creating fist-sized cracks in the surface of the field. It is only
natural to expect that adding water, kneading, and drying
would cause cracks to form.

But the ground at the corners of the fields, over which the
caterpillar tractors don't pass, is soft, rich soil just like that

in my own fields. I explained to one farmer that his soil had
not always been poor. I told him that the soil had probably
grown leaner each time the tiller had turned it over. Mechanized farming that assaults the land with chemical fertilizers
and pesticides as well as machines increasingly depletes it.

Scientists today claim that livestock farming enriches the
land. However, the truth of the matter is that wherever you
look, this only depletes it. After talking with young people
from Australia, India, and elsewhere, I've come to the
conclusion that animal husbandry destroys the fertility of the
soil. The question is *why*?

The Spaniards were the first to raise cattle in America.
The land, which one would normally have expected to become richer as a result, has grown poorer instead. In cattle-
ranching, because all the animal wastes are returned directly
to the land, it may not seem possible that the land would
grow leaner, but the fact is that it does. This is because the
vegetation becomes less varied.

The application of modern farming methods further aggra-
vates the problem, creating a negative cycle. In America, land
is irrigated with sprinklers and forage crops grown there.
These crops are nourished with chemical fertilizers, then
harvested using large combines, and packaged and exported
throughout the world as cattle feed.

The livestock raised in Japan is not fattened on Japanese
feed. Cattle and hog farms receive their feed from America.
By exporting all this feed, America is draining her lands of
fertility. Those who don't know any better might be pardoned
for thinking that America's livestock farmers are well-off.
But that is just not the case. All they are doing is selling feed
crops grown by applying petroleum-based products. The soil
at their feet just keeps on growing poorer and poorer. Oh
sure, they're making money. But the steady impoverishment
of the soil means that they are actually practicing deficit
farming.

When the soil gives out and the fields are abandoned by

ranchers, fruit growers come in. They set up sprinklers on the weary land, spread chemical fertilizers, and grow plums, apricots, and oranges. This no longer is crop production under natural conditions; it is farming based on petroleum energy. Even the water, although sometimes drawn from nearby sources, often is piped in from hundreds of miles away. This water is sprayed over the fields with sprinklers, but when it evaporates it draws salt up from within the soil. This continuous deposition of salt in the topsoil eventually turns the irrigated land into a salt field.

Agriculture Run Amok

Before going to America, I had intended to talk of the sad plight of the Japanese farmer and ask America's farmers to refrain from exporting so much grain, produce, and livestock products. But when I arrived, I found things to be quite different from what I had expected. I saw all too well just how great are the hardships of the American farmer. The crops he grows are not raised by the forces of nature; they are processed with petroleum energy. So there is nothing to be gained from being a farmer. The trading companies and middlemen like Sunkist make large profits by exporting fruit juices to Japan, but the farmers themselves merely practice crude farming methods with a very humble spirit. They live a plain, simple life, eating food of poor quality. Because they use modern machinery, pesticides, and airplanes, what they practice may appear to be modern agriculture, but it is in fact extremely crude, primitive agriculture.

On top of it all, they rely almost entirely on monoculture. Nothing but corn is grown in the midwest corn belt. That is all one's parents grew and that is all one's grandchildren will grow too—just corn. State after state is blanketed almost entirely in nothing but corn. Going further east, the fields are all planted in soybeans. It may sound ridiculous, but it is true; farmers grow hundreds and hundreds of acres of nothing but soybeans. That is absolutely all they grow. Proceeding east again, this time all we see are fields of wheat. These grain farmers grow almost no vegetables for home consumption. Since they are not self-sufficient, circumstances are tight. They plow fields a hundred times larger but live a more meager and deprived existence than the Japanese farmer on two or three acres. What's more the crops they grow do not receive the blessings of nature.

The reason America's farmers are so poorly off is that they have a severely disturbing influence on nature. And the ultimate cause for this, in turn, is the meat-based diet of Americans. The settlers and immigrants who came to America from the British Isles, France, Spain, and other European countries were all meat-eaters. Livestock farming for meat consumption began two or three hundred years ago when these lands were first settled. I believe that this has totally unbalanced American land. America has established an agriculture that does not grow the staff of human life, but caters instead to the hogs and cattle. I told people in America: "This country does not practice farming that sees after the needs of the land, does it?"

While in California, I had the good fortune of spending one week in the primeval forest at French Meadow. Here, dressed in a navy blue *jinbei* (see photograph on page 194), I spoke happily and at times passionately about natural farming and the principle that all is unnecessary to a group of over one hundred people, with towering trees and rocks at my back. I was deeply moved by a farewell campfire held for me on my last evening at the camp. I realized then that even I had been of some help.

Leaving French Meadow, we descended to the broad Californian plain below, then headed west to a commune where more than twenty young people from several countries were attempting to clear and cultivate a meadow in the Upper lake hills. I racked my brains for some way to overcome the problem of the solid cover of yellow foxtail grass. Then one night, beneath the stars, an excellent method for getting rid of this problem grass suddenly occurred to me. My heart secretly danced with joy. California's summer grasses had not withered at all; they were simply summer-dormant. I realized that these had to be awakened from their dormancy. This was because I had become convinced that the ambitious idea of blanketing the scorching Californian sands with green vegetation was more than just an idle dream. The next morning, I pledged with the youths at the camp to turn California into a lush, green land and make the rains fall here once more. We immediately set to work.

America's Pine Trees Are Dying Too ─────────

Earlier, on my way to the French Meadow camp. I had noticed that the pines growing on the mountains in California are dying too, just as they are in Japan. Actually, I found almost all of the pines in California already dead. It appeared as if they were hit by blight about ten years earlier than in Japan.

Although the varieties of pine differ, these trees have been dying in exactly the same way as in Japan. First one tree succumbs, then the following year several dozen trees in the immediate vicinity are afflicted. The initial symptoms are also identical. I concluded that the causes are the same in both countries.*

Once, while driving along a road, in the space of just one hour we ran into about twenty trucks carrying logs cut from the mountain. I called these "lumber hearses" and shared laughs with the Americans driving them, but it was a sobering sight. When I went to take a look at some land where lumber had been culled, I found that an area which had been clear-cut several years before had turned essentially to desert. Areas harvested of lumber are not being reforested; the land is often simply allowed to run to waste.

The pines are cut down because they have already died. That lumber is shipped over to Japan. For some time I had been saying that the putrefactive molds detected in dying Japanese pines were not native to Japan. Then, while in the U.S., I learned that these same molds are present in American trees.

---

* See pages 176–8 for a discussion of these causes.

I had begun to look into the problem of pine rot in the U.S. when I had the good fortune to meet the director of the state forestry bureau and discuss this and related matters with him. I told him that California seemed to have almost no trees left to export and suggested that it might be better to export *matsutake* instead. He was astonished when I said that a single *matsutake* mushroom fetched a higher price than a large log. The director introduced me to a number of university professors. From talking with them, I learned that American scholars have a different theory on the causes of pine rot. They attribute the disease to jet planes and drought. It seemed to me as if the net cast by American researchers is too coarse while that cast by Japanese researchers is too fine. As a result, neither are able to catch the fish they are after: a satisfactory explanation for pine rot.

America Is Drying Up ────────────────

From an airplane, it is clear that the third of the American continent east and west of the Rockies is a total wasteland, a reddish-brown desert. The image it brought to mind was of a flaming inferno. At 35,000 feet, what you see are circular farms measuring a half-mile across irrigated by huge, center-pivot sprinklers. These are very much like oases in the middle of a desert. And would you believe that America has 80,000 of these? Incredible.

Going further east, there is no doubting that the land along the shores of the Mississippi River is a grain belt. But from the sky the pattern of monoculture is all too clear. Fields of corn, wheat, and soybeans spread out below as far as the eye can see, leaving a vivid impression of the great expanse of the American "bread basket." The land is divided in checkerboard fashion into large, modern farms, but these are fields of death that have suffered the relentless injuries of heavy machinery, chemical fertilizers, and pesticides.

The soil has been pulverized by large tractors and the organic matter consumed and exhausted. Few microbes exist in the soil and when there is rainfall, the topsoil washes away at an alarming rate of perhaps ten or more tons per acre every year. Most U.S. farmland is hilly and rolling. There are no terraced rice fields or reservoirs as in Japan, so the soil just continues to wash out. When the loose soil dries, it is carried off by the wind as dust and scattered. And, as I have already mentioned, irrigation with those enormous sprinklers brings salt up to the ground surface, where it accumulates.

The land lacks the fertility to produce crops continuously year after year, so each year from one-third to one-half of the fields are left fallow. That is why, from the air, the land shows an alternating pattern of green and brown stripes.

Modern farming stresses labor productivity and efficiency. An inevitable outcome of this is the continuous monocropping of wheat or corn by large farming operations. This depletes soil fertility, setting into motion a negative cycle that becomes a basic cause for economic failure as well. Modern scientific farming amounts to no more than essentially slash-and-burn agriculture in total disregard of nature. The result is a decline in soil fertility, with the yield ratio (the ratio of energy yield to energy input) halving every ten years. Today, only one half of the energy poured into the land is recovered. To harvest one calorie of food energy, farmers are putting two into the land. Far from raising productivity, American agriculture can be better characterized as an industry striving to curtail declines in production.

Farmers who work farms covering many hundreds of acres on the shores of the Mississippi, America's richest grain belt, are unable to live as well and earn as much as Japanese farmers working just three or four acres. Faced with rising debts and deepening poverty, they are leaving the land. The fundamental cause behind this unfortunate trend is the destruction of the American continent that had its start in erroneous farming practices. This may very well be a just and proper retribution for those who thought they could put nature to work for them without fearing God and nature.

Surprised by this unexpected sight of farming in decay, as I traveled I spouted out that America was a land of artificial nature—a nation of poverty, and that it practiced a brand of agriculture that was primitive in its approach and methods. Americans were taken aback by this farmer from Japan talking everywhere of his astonishment at the state of the American continent. A write-up was done of me in the *East West Journal*, and I was even invited over to the United Nations.

After I returned to Japan, I saw several TV documentaries on the state of American agriculture. These confirmed that what I had said was not totally off the mark.

The East Coast Forests

On the East Coast, the scenery while traveling through three or four states near New York City was just the opposite of what I had seen in California. We were met by an endless sea of trees. It was as if we were traveling through nothing but forests. Woods made up of about five types of trees such as birch, maple, and oak—all about the same height—stretched on and on without end.

In California, I had ranted about how the American wilderness had been destroyed and was being turned into a barren desert, but on the East Coast I found a sea of green and recanted, saying that this was more like the America I had expected to find. However, after walking about for a week or so, I sensed that something was wrong here too. It occurred to me that this land had suffered the ravages of livestock-based agriculture. The proof is that, beneath the trees growing there, the soil is poor and depleted.

The glaciers are said to be responsible for this, but the last ice age was ten thousand years ago. In Japan, two thousand years would be long enough for a layer of soil five feet thick to form. The fact that this hasn't happened on the East Coast in the U.S. and that fifty-year-old trees are only so big appears to me a certain indicator that the soil has not yet recovered. If everything had been left up to nature, recovery would surely have been much more rapid. I concluded that this soil had been ruined by man and that the nature which remained was just a shadow of the real thing.

I suspect that what happened was that the early settlers on the East Coast kept driving further west and taking over Indian lands because their own lands wore out from the grazing of cattle. The fields they abandoned were left alone as unsuitable for cultivation and in time forests of mixed trees grew up on these. Of course, this is only conjecture based on a stay of just 40 days, so I may be mistaken.

I spoke once for about an hour to some of the people who work at Erewhon, the natural foods company Michio Kushi started in Boston. "If you did something with this mixed wood," I said, "you could become even richer than the Kushis." When they asked me to explain, I told them, "You could grow *Shiitake* mushrooms on sticks of sugar maple wood, for example." They broke out in laughter. This is, I believe, an infinite treasure-house. But no one is making use of it. Kushi himself said that he would entrust me with 550 acres in Ashburnham, Massachusetts, and asked me to do as I please there. That land is covered with the same mixed woods. If someone were to partially clear the land while using the cut trees to produce *shiitake*, they'd have every chance of succeeding.

The Bogus Nature

Boston and other cities in America have so many trees all over the place that it's hard to tell whether you're in a city or the woods. But when I climbed to the top of the 60-floor

John Hancock Tower in Boston and looked out across the city, I realized that there is not as much vegetation here as I had thought. Instead, I saw row upon row of buildings stretching out to the horizon below me.

Crossing town in a car, however, it looked as if one were passing through a forest. None of the trees lining the roads in the U.S. are pruned. Not a single branch is broken. The trees are left strictly alone. People in America do not fuss much with the shape of a tree by breaking and cutting off this and that branch. In this sense at least, they appear to be acutely conscious of protecting nature. That is why they just leave the trees alone to grow. In Japan, if a branch gets in the way, one promptly cuts it down.

But these trees do not seem to be native to the land. They were planted there, which means that the oldest date back no more than about two hundred years.

I was invited to lecture at a macrobiotics seminar held on the spacious campus at Amherst College in western Massachusetts. At the seminar, I addressed the following topic: "Nature has been destroyed in America. With nature dead, what thought and ideology do the people living there hold?"

I personally believe that without nature around, true thought cannot emerge. Everyone seems to believe that human thought and emotions are products of the human mind, but I think otherwise. The question I posed was, from where do human thought and emotions arise? We look at a flower and call it beautiful. We say that today is warm or cold, that something is interesting or dull. We call an experience pleasant or unpleasant and claim to be sad or lonely. From where do these simple emotions arise? In America people say they come from the head, while in Japan they are said to come from the heart. So the utterance, "That flower is beautiful," comes from either the head or the heart.

When we say it is "cool" outdoors, why is it cool? The scientist may tell us that the air is cool when the temperature falls below so many degrees, but he is just giving us a scientific explanation. I believe that this is something which arises naturally, from nature itself. A fresh breeze blows that makes us feel cool, that is all.

When people see a green tree, they all think that green trees are beautiful. Trees leave a sense of peace. When the wind ripples the surface of the water, the spirit becomes restless. Go to the mountains, and a sense of the mountains arises. Travel to a lake, and one feels the spirit of the water. These emotions all arise from nature. Go somewhere where nature has been disturbed and I doubt that anything but disturbed emotions will arise.

The road from San Francisco to Sacramento passes through an area that is now desert. The people of Sacramento appear to dearly love nature in their oasis of green. They take good care of the trees lining the roads everywhere. I found the same to be true in Boston. But are Americans who take such good care of this spurious green growth feeling the emotions that arise from cherishing real nature? Americans seem to be much more advanced in their desire to protect nature than the Japanese, but hasn't this arisen from a keen sense of regret at their loss of nature?

You know what I felt when I saw that lawn on the college campus? I didn't see a single butterfly about. There were no worms or ants in sight. This was certainly not the green of nature. Why, this was nothing other than a nature designed for the pleasure and convenience of man. People think that by protecting this they are protecting nature. If what they wish to protect is just a pale imitation of nature, then can their desire to preserve nature indeed be genuine and true?

At that seminar, I talked about why it was that the thinking of Americans led them to create such an artificial nature and allowed them to feel content with it. To me, as a Japanese, that lawn looks unnatural. There is no doubting its beauty, but it cannot satisfy someone who is Japanese. It doesn't put one in the mood to practice the tea ceremony or arrange

flowers; it leaves one unsettled instead. I told my listeners
that I felt as if it were impossible to feel truly in harmony
with nature there.

My talk proceeded to the question of which is more real:
the inability of Japanese people to feel content in a simple,
flat, geometrically symmetric park, or the ability of Americans to feel content in a green environment created by man.

Returning the courtesy paid by William Smith Clark,* I
got a bit carried away telling the young people there, "Who
needs learning if it fails to show you that the vegetation on
this campus is artificial? Youth of America, become aroused!
Don't you care whether the nature on your continent is transformed into a falsehood?"

Can American Agriculture Be Revived? —————————

What I'm most concerned about is which road America will
choose to take from now. Will it back modern farming
methods that make use of the bioengineering technology
riding at the vanguard of the next industrial revolution? Or
will it make a 180-degree turn, opting for a way of farming
that returns to nature?

Although I believe that not even one chance in a thousand
exists that America will opt for a method of farming that
returns to nature, in 1982 President Reagan ordered thirty
agricultural scientists to run a fact-finding survey on the
current state of organic and natural farming throughout the
world. I received a copy of the report, which concluded with
the statement: "The extent to which these methods are being
practiced today is extremely limited, but they deserve close
attention." I wonder what Reagan's purpose was in com-
missioning this survey.

---

* The educator and agricultural advisor to Japan who, upon leaving
  Hokkaido in 1877, told his students: "Boys, be ambitious!" Clark was
  an alumnus of Amherst and later taught there.

In any case, I believe that the only way to avoid the whole-sale destruction of American land is a complete reversal in current methods of farming. There is some room for hope as evidenced in the growth of interest among Americans in Oriental methods of farming, which treat the earth with reverence and care. This is an outgrowth of disaffection with modern agriculture. Oriental farming practices that attach importance to compost and manure were brought many years ago to France. Robert Rodale, chairman of Rodale Press, has run ambitious experiments on these practices in America, where they became known collectively as organic farming. Rodale Press is a publisher that serves the role of inspirational and educational leader for organic farming in America. It published my first book, *The One-Straw Revolution*, in 1978. When I visited the headquarters at Emmaus, Pennsylvania in 1979, the folks there were running large-scale tests on natural farming at the company's experimental farms. The results of those tests may hold the key to the future spread of natural farming.

At this point, however, natural farming has just barely made a start in America and Europe. Most people have only a vague idea of it as some new way of farming that is an extension of organic farming. But a growing number of individuals have begun to notice the fundamental differences in thinking and approach between these methods, and realize that a strict distinction will have to be made between the two in the future.

Organic farming, as practiced in America and Europe, is a take-off on the compost and manure-based farming practiced a while back in Japan. In many cases, people have applied these methods to grow pesticide-free vegetables in the suburbs, but a lot of groping is still going on. A great deal of effort is being devoted to devising new techniques for preparing compost, for the use of natural insect predators that eliminate the need for pesticides, and for companion planting, but organic farming is hardly off the ground yet. In fact, advocates have sensed the limits of this method on large farms and have been casting about for other methods to supercede organic farming.

One thing that surprised me during my visit to the U.S. was the great popularity of natural diet everywhere I went. In what could be the start of something big, natural farms of perhaps one or two acres are being established on the outskirts of cities to meet the large demand for health foods. These turn out relatively easy-to-grow crops such as tomatoes, eggplants, squash, melons, and beans. To a Japanese farmer, the method of cultivation is crude. Pesticides are not used so the produce comes in all shapes and colors, but customers drive up in their fancy, high-priced cars, buy freshly picked vegetables arranged in the stall at the entrance to the farm, and drive happily back home. Those wishing to enter the fields and pick the produce themselves are charged special prices.

Natural farming in America has only just begun. Most often, the incentive comes from an interest in natural diet. But attempts at natural farming are also being made at farms run by back-to-nature groups and at Zen centers and the like. There are reportedly more than 250 Zen centers in the U.S., most of which operate farms. Rather than viewing these as outposts for Eastern thought, they may be thought of as Zen farms that have totally merged with the American scene. On weekends, people from the cities take part freely in the activities at these centers, irrespective of religion and sect. They enjoy farm work and Zen training as a means of achieving serenity. These are places where Zen and agriculture are united and an effort made to create a cheerful and happy utopia, so natural farming is readily adopted and put into practice.

I was taken by a priest of the Buddhist Nichiren sect to visit a Korean Zen temple in southern New York that had begun natural farming. There I found two Japanese sisters of about junior high school age. They were doing their best to grow tomatoes using natural farming methods. When I talked with them, I learned that their mother, who lives in Kobe, had sent them here, saying that Japanese high schools are decaying. I was astounded and impressed. The simple, rustic life that Japanese people have forgotten is still to be found in America's countryside.

So I found America to be an interesting country. Wherever I went, I met young Japanese people who, unable to practice natural farming in Japan, had gone to America to try it there. Reversals in thinking and farming methods can arise in unexpected places.

Of course, the problem that lies ahead is whether or not natural farming is feasible in an agriculture system that relies so heavily on the cultivation of enormously large fields. Here and there I saw how people who have turned from the blind alley of scientific agriculture and are attempting to make a workable solution of organic farming are baffled by the problem of how to handle compost and manure. With the help of machinery, they are able to feed hay and fodder to cattle and even to collect the wastes from these animals, but they generally have trouble when it comes to transporting that compost and manure out to the fields and spreading it over large acreages. Seeing the large piles of compost and manure left exposed everywhere on the fields even at the Rodale farm, I said, "It's more important to create a grassy cover of green manure than to spend so much effort on composting." People showed a great deal of interest in this idea. I felt that once the theory and practice underlying my belief that composting is unnecessary take hold, it may be possible to convert even large farms from organic farming to natural farming.

By contrast, I was surprised to find that converting wet-rice cultivation in California over to natural farming was not always as difficult as I had anticipated. When I spoke for a half-day with one farmer, Harlan Lundberg, who has a 7,500-acre farm in the Chico area, he grew excited and right then and there decided to switch some ot his fields over to natural farming. Needless to say, I was the one who was surprised. California's climate is suited to the cultivation of rice by natural farming. One can start right away by abandoning the tractor and changing the variety of rice and barley planted. Moreover, because the yields go up, if the will is there, then there is every possibility that natural farming could take hold and spread throughout America's great farming belt. As a matter of fact, with the American temperament, it is quite possible that natural farming will succeed earlier in America than in Japan.

As I watch those visitors to my farm who, turning their backs on English civilization leave for Australia saying that they are going to begin natural farming, I get the sense that, although the world seems to remain forever the same, it is quietly moving in a very big way.

However, it doesn't look to me as if natural farming has really gotten underway yet. Until the spirit of natural farming and Eastern philosophy that rejects Western philosophy is fully understood, any effort made is just hollow mimicry. Natural farming is more than just a means for growing health foods or enriching the earth.

Even if one is unaware of the pitfalls of modern farming, the changeover to natural farming involves a sweeping, Copernican transformation. It is not something that can be accomplished overnight.

The Machine Culture Is Doomed

As I have already pointed out, nature in America is not natural at all. This is a human-centered society and way of thinking that has its beginnings in a pact by Western philosophy with God. Americans are a meat-eating people who have farmed for meat consumption, triggering a negative cycle that destroyed nature and laid the foundation for a machine-based culture.

Why has American agriculture and nature gotten into such a fix? Up until now, Americans have moved toward enlargement and expansion, preferring large over small, rich over poor. Politics and the economy too have all raced headlong in the direction of expansion. This, after all, is what modern civilization and development are all about. Yet all this has really been just a fall from grace into the inferno.

Such expansion has reached a dead end in our machine culture and in urban cultures such as New York City. Those trapped within are all trying to escape. I myself spent several days in New York and even walked about some at night. I found none of the individuals that I met frightful or menacing, whether in Harlem or elsewhere. They all seemed to be very good people. I even thought that, if anything, it was the blacks who were able to laugh from the heart. There is an area frequented by drunks right at the center of that enormous city.* Looking at the drunks there during the daytime, I felt that these are truly cheerful faces. But when I looked at the faces of the smart and clever, those living affluent lives, none of them bore an expression of contentment. All had a tragic, cornered look on their faces. I believe that was a direct indication that this culture has come to a standstill.

* The Bowery, in lower Manhattan.

It is true that this city is a pit of crime, a world of civilization in despair. The first place to go under the moment the oil stops flowing will be New York. People are trying to break free of that situation.

When I declared that the nature found in California and on the East Coast was just a spurious nature, not real nature at all, I was told, "That may very well be so, but it all goes without saying. After all, we invited you over because we're trying to change things."

Yes, indeed, there is a willingness here to accept and try natural farming. No matter how much nature has been harmed, America is a vast land with infinite possibilities. It is time to turn priorities around, to choose the small over the big, to decide not to develop rather than to develop. Isn't it enough just to be alive?

"All I've done is to strive constantly to do less and less in order to eventually not have to do anything," I told my audience in Amherst. "I've spent close to fifty years farming in a way that keeps me looking out for things I don't have to do rather than to find new things to do.

"People say that there is a purpose to life and talk about what it is that makes life worth living, but life has no purpose. This I realized fifty years ago, and I saw also that there never was anything that we had to do. Such ideas are just notions that man dreamt up himself.

"People set tentative goals under the illusion that achieving these will make them happy and well-off. They think that doing nothing is the most boring and empty life that could be lived. But no, it's just the other way around. Doing nothing, living life without a goal, taking a quiet nap when it suits you—that is the door to the most pleasant of worlds.

"The only road for man is to do nothing. If I were to carry out a social movement of some sort, this would have to be one that consisted of doing nothing. If everyone were to do nothing, the world would become a place of peace and plenty. Nothing more would have to be said."

This went over very well at Amherst. On the campus at Berkeley, people in the audience even came out and said that when they returned home they would become farmers and practice natural farming.

Food as a Strategic Weapon ─────────────────────────

America is a powerful and affluent country, but it is also a country in great danger. Depending on how it uses the food it produces, such a large country producing so much food is capable of saving the world or of throwing the world into chaos and even destroying it.

Today, this food is used as a strategic weapon. The country really has little choice since it is food produced with oil. Transported elsewhere and sold for profit, this food serves as a pillar of the nation. This is why the Carter administration kept after Japan to buy American oranges and wheat, and why, when Japan said it would ship its surplus rice over to Vietnam, the U.S. State Department roared out in protest. The U.S. told Japan to cease and desist because if Japan were to send its surplus rice to Southeast Asia, America would not be able to sell its own grain. A stern word of warning from the U.S. Government was all it took to send tremors of fear through the Japanese Ministry of Agriculture, Forestry and Fisheries.

Food today is a strategic weapon of the U.S. What would happen if this were changed and American farmers began to practice the type of organic farming traditional to Japan, or even better still, natural farming? Rather than using large fields to grow food for export to other countries, farmers would be able to produce abundant food on small acreages and live a rich, meaningful life. Farmers working many-thousand-acre farms in central California grow rice one year, leave the land fallow the next simply to get rid of unwanted millet, and raise summer wheat the third year, meaning that they grow only one crop of rice every three years. By growing rice every year and raising this back-to-back with a second crop of wheat, it would be possible for them to increase starch production three-fold. When I told one rice farmer that if the state government put its mind to it, in three years the state could be growing as much rice as all of Japan, he jumped up and said, "Why that's incredible. This is a revolution."

The land gets plenty of sunlight and has enough water. Some people in Japan have even tried to keep me from saying too much because they fear that if rice were grown on an even greater scale in California, that would be the end of Japan. They ask me if I shouldn't be more cautious about going over to the U.S. and in effect sparking a movement to increase rice production.

To be honest, I thought the same way myself at first. I felt that with the unlimited resources available in California, if that state ever got serious about growing rice, Japanese farmers didn't stand a ghost of a chance. But then I realized that this wasn't so at all. I sensed that the root of the problem was the poverty of the American farmer. If American farmers had a better diet and lived a richer, happier life, then they would have no need to export crops to other countries. The real reason why the American farmer has to sell his crops to other countries is because he is poor.

When I visited the United Nations in New York upon the invitation of an American official working there, I told him. "America and it's farmers are not affluent at all. Actually this is a poor nation. The food tastes bad, the soil is impoverished, and there don't seem to be any resources here at all. No wonder the U.S. buys foreign oil and uses it to produce food which is then shipped abroad and used as a strategic weapon.

"This country appears to have come under the illusion that it can control the entire world. Try producing truly abundant, naturally blessed food that serves as a real source of life and have all the people of the nation eat a full and wholesome diet, and you'll see that your country never had any need to export food to other countries."

California's Sunkist Corporation is hurting mandarin orange production in Japan, but when I traveled around California and bought fruit directly from farmers at roadside stalls, I found that for one dollar you could buy a bagfull of oranges or three large melons. The farmers griped about how hard a time they had making ends meet, yet upon returning to Shikoku, I found that melons sold in the nearby city of Matsuyama were priced at 1,500 yen apiece. The American farmers growing fruit and vegetables for export are definitely not getting rich for their troubles. Why indeed would these farmers wish to put pressure on Japan?

It is not the American farmer who is leaning on Japan. People working for American export and merchandizing firms are the ones bringing this produce over. It is a small group of trading companies that is trying to wipe out the Japanese farmer. And people in the big cities are offering them a hand.

Consumers in the cities don't know how food is produced or the mechanism by which it is priced. They know nothing of American or even of Japanese farmers. All the consumer is concerned about is buying food that is inexpensive and good-tasting. There is no getting around it, both the Japanese consumer and the agricultural leadership in Japan have gone off the deep end. They point the finger of blame at someone else but are themselves accessories to the crime for they share the same view of things as the villains perpetrating this evil.

Everyone believes that as long as they can enjoy good-tasting food for cheap, it doesn't make a bit of difference whether they're eating fruit grown in America or Japan, whether it's American or Japanese rice that they eat every day. They don't seem to have noticed that this is a lunatic error.*

---

* I noticed, in 1979, that California rice sold throughout the U.S. was set at about half the price of Japanese rice, but the price of gasoline was also about half that in Japan.

People don't know what real plenty is; they have no idea what to grow where. The starting point of food production is the inseparability of the body from the earth; it is self-sufficiency. It should be readily apparent from the fact that large-scale monoculture and distribution is the cause of the poverty of the American diet that the notion of an international division of labor is a preposterous fallacy.

America today prides itself on its advanced culture, and appears frantic in its effort to drive forcefully ahead with its dual arsenal—military weapons and food—to sustain and assure the continued prosperity of this culture. However, a careful look at this strategy shows that it is full of holes and contradictions; it is, in fact, collapsing.

At one university in Boston, radiation has leaked out to the exterior wall of a cylindrical building housing an experimental nuclear reactor. I was horrified when I looked through a wire-mesh fence at the grass growing along that wall.

I remember meeting twenty young people who had fled from Three-Mile Island with twenty turkeys (three of the birds later died of radiation exposure). I encouraged them to try natural farming: "You'd perform a greater service to society by setting up a self-sufficient life for yourselves with natural farming in these hills and demonstrating just how enjoyable life can be without electrical energy than by taking part in a movement against nuclear power."

I was warmly received at an Indian reservation and learned for the first time how soundly and peacefully it is possible to sleep under a roof through which one can see the starry sky.

I believe that the run-down, profiteering taxicabs that barrel suicidally down the streets of New York are symbolic of the frightful decay of American culture. From the indigence of farmers in the country's farming belts and the poverty of their diet, I was able to see that the errors of farming methods that arose from the illusions of Western philosophy are destroying nature, annihilating the land, and even wiping out whole peoples. What I saw in the U.S. confirmed for me what I had already suspected: aberrations in agriculture create aberrations in urban culture. Who is going to strike back at the philosophy of the U.S. government, which persists in the belief that it can lead the world through its two-pronged nuclear and food strategy?

The way I see it, never has there been a more fitting time in which to learn from the ancient ways of the American Indians. Placing the last ray of hope on a revival of the Great Spirit of Mother Nature—the soul of the American continent, I started back for home.

This is true of America, and it is true also of Japan. What is most sad as I look back is to see Japan following in the footsteps of America.

The Natural Foods Boom

Natural Diet Takes Hold in the U.S. ─────────

The modern macrobiotic movement was founded in Japan by
George Ohsawa. Today, his students are working throughout
the world to spread natural diet. The two leaders of the
health-through-macrobiotics movement in America are Michio
Kushi in Boston and Herman Aihara in California. It was
through their good graces that I was able to tour America
in 1979.

In the 1960s, Kushi coined the term "natural foods" to
distinguish whole, unprocessed food from "health food,"
consisting largely of vitamins and dietary supplements. His
company, Erewhon, popularized organically grown grains,
beans, vegetables, and fruits, as well as seaweeds, tofu,
tempeh, and other high-quality foods in North America.
When I visited, I was astonished to see natural foods such as
miso, soy sauce, and brown rice being exported from large
plants and warehouses to the entire world.

Of course, Kushi and his family went through times of
hardship. "Fifteen years ago, I and three of the children were
packing miso into bags for sale in a tiny room," his wife
Aveline told me. In 1981, after natural foods spread around
the world, the Kushis left food production and focused their
attention on education. Claiming that conventional schools
today are no good, Kushi is working to establish an international college based on principles of natural order in order
to revolutionize education. He has institutes in several countries teaching macrobiotic cooking, philosophy, and medicine,
as well as spiritual development and the reconstruction of
society in a more peaceful direction.

Although natural foods are enjoying something of a boom in Japan too, this can't begin to compare with the scope of the movement in America. What is happening in essence is that natural diet is being imported back into Japan, where much of it originated. During his lifetime, Ohsawa was virtually ignored by the Japanese. In Japan, natural diet has gained some popularity with the concern over pollution, but the core of its adherents are people in poor health who pursue this largely as a form of therapy. In the West, the starting point is different. There, a radical change has occurred in the basic notions about food as people have been won over by the principles underlying natural diet, which is a direct outgrowth of Eastern thought. This has given the movement a solid foundation there, which is why, once the natural foods industry became established, the movement grew with such incredible force.

While traveling through the American countryside, I got a feeling that the day will come when rice will take on a much greater importance there. Many American farmers are saying, "I want to grow rice in my fields. Can it be done?" They believe that rice is the most promising crop partly, of course, because of its profitability.

The attitude in Japan is clear, what with the government's policy of cutting rice acreage and the complaint that there is too much rice, that "we don't want rice." Just the opposite is true in the U.S. Americans are shifting in their dietary inclinations from meat consumption to Eastern-style vegetarianism. This is especially apparent now as the natural diet movement has started to come into its own and the custom of eating rice, and in particular eating and carefully chewing brown rice, has become established through natural diet. I get the feeling that the American palate has changed where taste is concerned.

Up until now, Westerners have eaten almost exclusively meat and dairy products. With their concern today over cholesterol levels, cancer, and obesity, their physique may have improved, but the people themselves have become odd. And that in turn has made the entire culture a bit strange.

Having noticed the danger to civilization and the disruption of their own body and mind, people in the West have begun to have second thoughts. They realize, upon reexamination, that the illness that afflicts their body and mind originates in their diet. What should they do to correct their diet? After deliberating, the solution they hit upon is an Oriental natural diet. But this natural diet is nothing special; the traditional Japanese diet would have done just fine. What all this means it that the diet eaten by Japanese farmers of old is the diet in greatest need by Americans today.

A few years ago, a young American came to Japan, studied tofu, and wrote a book about it.* His book contains five hundred recipes for tofu and is today a best seller in the U.S. Because the soybeans themselves are grown in the U.S. and because, in spite of their inventiveness, Americans remain faithful to the basic methods of preparation rather than being exclusively concerned with tastiness, handmade tofu of excellent quality is made and enjoyed throughout the country. I'll have to admit that, even to the Japanese palate, tofu made in the U.S. is better-tasting than that made in Japan. This is surely why Americans think of tofu as such a good food.

Many American supermarkets today have a shelf stacked with 5-pound and 25-pound bags of rice. Next to the rice are bottles of *amazake*, a sweet, fermented rice beverage. In Japan, one finds coke bottles and cans wherever one turns, while in U.S. health food stores, one sees instead large displays of *amazake*. Above the *amazake* will be a shelf of popped rice cakes. There is so much of it that it would almost seem as if this is being used as a bread substitute. In any case, rice and rice products take up a large part of an aisle.

People in the U.S. buy these foods and eat them because, since the Japanese food boom among Americans arose from the macrobiotic movement, people in the States think of foods that were essentially a part of the traditional Japanese

---

* *The Book of Tofu* by William Shurtleff.

diet as "delicious." They know the true meaning of the word delicious. One of the reasons Americans have come to regard Japanese food as so good may be that European cooking is not very appetizing. Not only is there little variety in the materials used, country cooking in America and Europe is not fully developed. There is no sense at all to the dinner table. It is all very crude and careless, the only principle at work being that of nutritional science. Meals are prepared with the belief that the body is sustained by three nutritional elements. Cooking in the West thus follows the same reasoning used to raise hogs on synthetic feed. Few think of seasoning their food and eating something delicious. No one eats anything good, so even simple Japanese country cooking becomes a welcome treat.

Oriental food has thus been well received and restaurants serving Japanese food are popping up everywhere in America. Of course, this boom consists largely of sushi shops, tempura shops, and the like. Moreover, the sushi and tempura they serve is second- or third-rate by Tokyo standards. Still, the customers are crowding the shops so they must like what they are getting. Because they are used to bad-tasting food, such modest Japanese fare appears to be a major treat. So one reason for the popularity of Japanese cooking is that Americans have begun to regain a true sense of taste. But another reason is the fact that this food is good for the body. I believe this explains the accelerating boom in Japanese food. People in the States see Chinese cooking as being delicious, and Japanese cooking as being good for you.

While I am on the subject, I might add that Oriental foods are also enjoying growing popularity in Europe too. In many West European countries, one can find miso and soy sauce —sold as "shoyu" or "tamari"—even in country stores. And in restaurants, a bottle of soy sauce is generally put out on the table together with the other sauces. So Japanese food can be found almost anywhere today. Rice, however, is grown only in France and Italy, from which it is shipped and sold in countries further north such as Great Britain, Belgium, and Holland. From what I hear, the bulk of this rice is brown rice grown by natural farming. Naturally grown brown rice from the Milan area is especially prized. Rice grown scientifically is regarded today as second-rate food.

I already pointed out how I believe that, as meat-eaters, Westerners have suffered a numbing of the palate, making them incapable of perceiving true taste and thus unable to appreciate subtle flavors. But when they begin to eat brown rice and switch to vegetarianism, their sense of taste also returns. That is the way I see it. The flavors of nature are being revived by adherents of natural diet in Europe and America as their sense of taste becomes sensitive once again.

I felt this very strongly when Herman Aihara, leader of the macrobiotic movement on the West Coast in America, visited my farm once a couple of years ago with about twenty students during a tour to study the current state of natural diet in Japan. His followers were not very impressed with what they ate at a first-class Japanese restaurant. They came to stay overnight the following day at the huts in my orchard. The next morning, I prepared a gruel of brown rice, threw in some greens picked nearby, and because there weren't enough bowls to go around, made some cups by cutting cylindrical sections from overgrown bamboo shoots. When we served this gruel mixture in the improvised cups and everyone sat down to have breakfast on the clover growing in the orchard, my visitors said, without the least hint of flattery, "This gruel is great!" When I heard this, I realized that the American palate has changed.

The Japanese, meanwhile have turned increasingly to a bread- and meat-based diet, distorting their sense of taste.

A Change Has Come Over Westerners ──────────

Descartes said: "I think. Therefore I am." This reflects the Western viewpoint up until now. What it says is that "If I did not exist, there would be no nature"; "nature" exists because this self-reflective "I" exists. That is why people in the West have felt that nature may be used and refashioned as man pleases to serve him. This notion has been the starting point for the development of sciences that serve man. Using such science, Westerners have controlled the world and other races. But people have begun to realize today that there is something wrong with this; they've started to notice the fallacies of Western philosophy. They are saying, "We see that something's wrong, but we have no idea which way to go from here."

What I think is happening, very briefly, is that Westerners have started to question Western philosophy and the road taken by Christianity. They have begun to notice that new possibilities exist in Eastern philosophy and Buddhism. They are not saying that Eastern philosophy and Buddhism are good, or that they will abandon their Christian beliefs for the Buddhist faith. They aren't going that far but they have noticed that there is something good there.

Impressed by the fact that Westerners had at one time conquered the world, the Japanese thought their exploits dashing and were convinced that this was progress. As a result, the Japanese too have become clever, their diet rich, and their physical stature larger. Why, they have even begun to practice sports and are enjoying a high level of material affluence. Having come so far, my countrymen have even begun to think that they may be able to conquer the world and are starting to feel proud and mighty. But in the West, people see things differently. They are saying, "All the Japanese have done is to travel in thirty or forty years the road that took us a century or two to build. Perhaps they have caught up and even passed us. But we don't envy them for it because that road leads nowhere." Asked where the future lies, they respond, "The Japanese have all but forgotten Eastern philosophy. But we will probably go in, mine this forgotten wisdom, and use it to progress in a new direction."

This is more than wishful thinking on the part of Westerners. I can sense this possibility just from observing the many people who come to Japan to learn about Zen and natural diet. Many young foreigners who visit my farm will come over after stopping by a Zen temple such as Eihei-ji or practicing Zen meditation somewhere. They are able to clearly and unequivocally say, "I don't understand." On the other hand, after doing *zazen* for a short while, Japanese will put on a knowing air and say that they understand Zen. They arrive at a point where they *feel* they understand, and so they stop meditating. Westerners, on the other hand, are not afraid to admit that they do not understand. They will say, "I came and tried meditating at a Zen temple, but nothing came of it." What they mean is: "Nothing came of it and I don't understand it all, but since I do see that I've been traveling the wrong road up until now, I guess that, for better or worse, all I can do is to go on meditating. I realize that nothing comes of practicing zazen, but I don't have any intention of turning around and heading back for home. My only choice is to continue with my meditation." That may be a very small distinction but this is where the big difference with the Japanese lies.

Having noticed that the Western diet is no good, many in the West are switching to an Eastern diet. This is different than saying that one cannot tell whether an Eastern diet is good or not without trying it, so let's try it. The Japanese try a natural diet when their health gives out, but as soon as their condition improves a little, they return to their former diet and begin again to eat "delicious" food. They drift from right to left and back again. The Japanese, whether for better or worse, tend to do things halfheartedly. As for Westerners, once they have decided for themselves that something is no good, they reject it and never turn back. Westerners following a natural diet who come to visit my farm never, under any circumstances, eat fish or white rice. If they are on a brown rice diet, they will stick strictly to that. If they don't eat fish, they won't so much as eat a single small dried sardine. That is how singlemindedly dedicated they are. It is not a matter of good or bad—that is not their concern. Rather, once they've made up their minds, they stick to that decision. It is very simple and clear. I have a feeling that the day will come when this simplicity will be what counts. The Japanese have lost their simplicity and reason with their minds: "If A doesn't work out, then I'll try B next." When they discover that B too has its advantages and disadvantages, they arrive at a decision that is a compromise of both and, relying on their own judgment, elect to follow a third road. Unable to reject the first road or take the second road, they get lost on a different road altogether. As a result, everything ends up half-baked.

In contrast, people in the West are capable of choosing one path and sticking firmly to it. That is why, when I see that practically no one in Japan is practicing natural farming and compare this with the rapid increase over the past few years in the number of those practicing it abroad, I cannot help admiring the ability of Westerners to make decisions and follow them through. In a sense, whether you are talking of Zen, Buddhism, or Eastern philosophy, this is largely just plain mimicry, but Westerners have this wonderful simplicity of heart that allows them to pursue these knowing full well that what they are doing is little more than mere imitation.

Regardless of what people may say, if one adheres carefully to the sort of vegetarianism practiced by the Japanese and other Oriental people of the past, then the body also changes. Cosmetics have made Japanese women glamorous, but I suspect that they must lose the beauty of skin that Oriental people who kept the vegetarian diet of old had. Perhaps cosmetics are used to offset this loss of skin beauty. In the same way, there no longer are any true Japanese. They may

look Japanese on the surface, but their way of thinking has
become Westernized. Even the physique of the Japanese is
becoming more Western. Perhaps the shape of the body is
more stylish and the build more imposing, but the Westernized
diet has destroyed the luster and beauty of the skin.

While we in Japan have been looking over to the West and
becoming Westernized, Westerners have been turning this way
and adopting our ways. Although they may have little more
than a cursory understanding of Buddhist thought, as they
pursue a vegetarian diet, their body comes to resemble that of
Oriental people and their behavior becomes like that of the
Japanese. Although they claim not to understand Buddhist
thought and philosophy, they are living a Buddhist life.

Even if a priest is unable to understand Buddha's thinking,
he will be satisfied with imitating his way of life; that is the
way a priest lives, as I understand it: vegetarianism and celi-
bacy. Of course, that in itself is hardly a beginning. But I
think it is enough to follow the Buddha while copying his way
of life. When I tell them that Lao Tzu said, "It is enough to
live a natural, 'do-nothing' life," Westerners say to me: "No
amount of thought is going to help me know whether 'do-
nothing' nature is good or bad. That's why I've got to try
it first and see." It is the Japanese today who say, without
even giving it a try, "That looks duller than city life, so I've
given up the thought of trying it."

To the question of when one should venture such a life, I
believe this becomes possible when one's body changes. I
think that when the body says, "I want to eat meat and fish;
I want to watch TV and listen to the radio," bringing about
a change in that person's thinking will be an impossibility.
But when one lives in the country and one's diet and body
changes, everything becomes clear of itself.

I get a lot of people who think they want to become farmers
and come stay at the huts in my orchard to get a feel for the
life. But what bothers me with many of them is that although
they may think that way, inside they haven't really made up

their minds. They talk big about how they have come because "life in the cities is meaningless," but all they want to do is give the life of a farmer a try. If it doesn't suit them, they soon abandon it. The only ones who are able to live for any length of time on the mountain are those who are sick or in total despair. People dedicated to a natural diet or vegetarianism, who find the greens growing by the huts in the orchard delicious, also adapt well. But those who come with their big-city mind and attitude to enjoy a bit of nature in the hills and get a nice tan, and who think of returning home while working in the field or orchard, are here only in body; their minds remain in Tokyo. This won't do. Both the body and mind must agree. It doesn't matter which you start off from, but both must be in accord. In a way, a diet of brown rice and vegetables means little in the city. Even when one lives in the city and resolves to stick just to brown rice and vegetables, sometimes the mind is off somewhere else. This just won't do. It means either the mind, the body, or the soul is absent. It is just like the farming youths recently who talk of taking up natural farming because ordinary agribusiness doesn't pay. Just why does someone practice natural farming anyway?

Only when the mind, body, and soul are in agreement can one make a proper start. But when talk gets around to where this start should be made, people find self-reform very difficult. First they have to be rapped on the head so it hurts, or they have to trip and fall down. When they see themselves in a bind, then finally they are able to act. That is the way things are. So people have to take a spill before they are ready to try.

The Specter of Food Scarcity

When I went abroad, I became intensely aware of just how wonderful Japanese farming is, how extraordinary the Japanese farmer, how precious the land that he protects, and how amazing the methods that have kept a land fertile for three thousand years. Nowhere else has there been a way of farming that has taken such good care of the land. Nowhere else have there been fields that produce a crop of rice each year while maintaining such a rich, dark soil. The food that was once grown on that dark soil was the best in the world. But that food has vanished and with it has rapidly disappeared the happiness of the Japanese farmer.

I am one of those who believe that we have no way of knowing what lies ahead; I am unable to prognosticate. I have never said this before, but my prediction, if I must give one, is that we cannot read even a split second into the future. Things could go either way depending on our approach. If we insist on feeling that everything is okay the way it is now, we'll end up at an impasse. Adherence to our modern, energy-wasteful scientific farming methods will only result in a further destruction of nature, lower yields, and a continued decline in the quality of our food. Of one thing I have no doubt. Because Japan imports 60–70 percent of its food, it will get hit first and hardest should a food crisis occur. But I wouldn't call the situation hopeless, since even this could go just about any way imaginable.

The first step taken should be to stop eating meat. If the Japanese stopped eating meat and ate the Japanese peasant diet of old, I think the country could be self-supporting even with twice the population it has now. This is because eating meat involves the consumption of more than seven times the calories as eating grain. In other words, we are spending seven times the amount of energy we need to, which is amazing extravagance. If everyone gave up eating meat and growing feed crops and took instead to eating cereals, tubers, beans and the like, that alone would make an enormous difference.

However, if people continue their fixation with meat because they feel it to be delicious, high in calories, and nourishing, the scarcity of food will accelerate. As people in the Japanese cities, won over by publicity claiming that "the Japanese still eat many times less meat than in the West," continue to eat more and more meat without being aware of their own immoderation, and as people in developing countries—particularly those to the south—take up meat-eating to keep pace with their rising standard of living, this will surely help precipitate a global food shortage.

It is thus futile to try and predict whether a food shortage will arise. Everything depends on what people eat, what their food staples are. So there is no need to worry so much about the future. The problem is, in which direction does man intend to go? Will he follow his cravings and eat whatever food he desires, or will he exercise self-discipline? How things turn out depends on what the housewives in the big cities decide, what people eat today, and what agricultural methods the farmer uses. There is a danger that, if nothing is done, things will grind to a standstill in five to ten years. Well . . . I think we ought to regard the situation twenty years hence as highly uncertain.

The global food situation is indeed a large problem. America and the United Nations are turning out quite a lot of statistics on this. When I met the head of the United Nations World Food Council, I received a large stack of literature. But, depending on one's viewpoint, this data could be interpreted in just about any way. Statistics are published and used to plan for the future, but the most important factor of all is the attitude of people. Everything turns on this. The true food crisis begins when people lose sight of the spirit of Christ's words: "Man does not eat by bread alone." Unless the true meaning of these words is renewed in people's minds today, there will be no avoiding the hell of human famine.

2

Europe
As I Saw It

Touring Europe in *Geta* and *Monpe*

In 1983, I went to Europe. I had been invited to lecture at several summer camps. A Greek fellow by the name of Panos and a young Italian woman named Miriam, both of whom had stayed and worked on my farm, offered to serve as my guides, so I decided to go. Altogether I spent fifty days over there. With these two taking care of everything for me, I was able to enjoy a pleasant and carefree voyage in *monpe* (baggy peasant work pants) and *geta* (Japanese clogs).

I traveled by car through five or six countries, enjoying the beautiful scenery and visiting major farms working to adopt a new way of farming. For the most part, I spoke to groups of people who had gathered to hear me. On a few occasions, I even spoke at large municipal auditoriums. I did a bit of sightseeing throughout my trip, and on long outings even stopped from time to time at roadside cafes for tea. Wherever I went, Westerners and Japanese I knew directly or indirectly were waiting for me so I suffered no inconveniences over language or food. Simultaneous interpretation into three to five languages was generally provided at the summer camp lectures.

The Sound of My Footsteps ━━━━━━━━━━━

My unusual attire was responsible for a number of unforeseen experiences. I had boarded the plane to Europe dressed comfortably. When we touched down briefly at Anchorage for a fuel stop, I got off to stretch my legs. As I was making my way through the shopping area at the airport there, a young Western girl working at one of the souvenir stands suddenly

yelled out in Japanese, "Welcome, Hanasaka-Jiisan!"* I wasn't
the only one taken aback. All those near me turned in surprise,
then we all broke out laughing. This gave me a chance to
speak with the people around me. Normally, people get onto
a plane acting uptight as if they were enemies. But when I'm
around, this helps loosen things up.

At the de Gaulle Airport, I was wandering around trying to
find the gate boarding for the plane to Switzerland when a
gendarme came up to me. "Wonderful!" he exclaimed at the
small man in strange clothing he'd found, and he promptly
took me to the gate himself. I just followed meekly along,
becoming some sort of a toy. People treated me all the more
kindly as a result, and this enabled me to approach them and
feel my way through situations.

As I walked, my *geta* rang out on the cobblestones. That is
when I discovered just how good a sound Europe's cobblestone
streets give off. Japan's roads are paved with asphalt. This
made me reflect on what type of road is good for walking.
Europe's streets, both in the large cities and small country
villages, are still paved with stone, as in the past.

I got a sense of the strong attachment of West Europeans
to stone pavement when I watched them repairing their roads.
They aren't out there working on asphalt roads with heavy
machinery. Workers wearing leather leggings dig up and replace the pavement stones one at a time. Laid down strategically long ago, town roads twist and wind and are almost
impassable by car. These are left unchanged because people
do not wish to destroy the stone pavement. As a result, no
straight roads run through the towns.

The top surface of a paving stone is 6 inches square, and
the stone runs about a foot deep. Carriage tracks and the
sound of horses hooves are said to be carved into these.
Because each stone is different, walking over the road creates
music with a rhythm all its own. Whether you like it or not,

---

* An allusion to the main character in a Japanese folk tale who was
dressed similarly.

as you walk, you constantly hear the sound of your own footsteps.

It occurred to me also that the sound of the Orient echoes within the *geta*. Whether this has a ringing cadence to it (ke-tang', ke-tong'), an even beat (clippety, cloppety), or a lonely "picky-pocky" depends on the weather and climate. And when one feels rushed inside as one walks, the *geta* call out: "What are you in such an awful hurry for?" The *geta* always give warning, whether you walk quickly, slowly, or absentmindedly.

Listening to the sound of my *geta* even put me in the mood for writing some bad haiku:

> Touring through Europe,
> My constant companion—
> The sound of *geta*,
>
> Ke-tang', ke-tong'.
> Listening to my footsteps—
> A *geta* journey.

The sound *geta* make is truly a sound of Eastern philosophy. It seemed special to me that not only was I able to read my own mood in the sound of my *geta*, I was able also to tell something about the national characters of people by their reaction to the *geta*. But in any country, it is the children who are the first to notice. They would stare intently at me from far up the road and continue watching until I had passed. Sometimes I even teased them playfully. But adults did not notice. Englishmen, Austrians, and Germans strutting along self-importantly, with their heads held high in a pompous manner were very slow to take notice of my *geta*. But the self-deprecating Italians, who claim to be the most ridiculed people of Europe, notice quickly. This is because of their low posture, which is apparent even from the way they walk. You can see this from how low they hold their heads and their physical build. The Dutch, who themselves wear clogs or sabots, were also quick to notice my *geta*.

I mentioned this to people, pointing out that everyone walks differently. "The English walk with their heads, the Germans with their shoulders, and the French with their chests. The Italians walk with their legs, swinging their hips and looking down as if to avoid being ridiculed by other Europeans." When someone added, "French women swing their breasts as they walk," we all had a good laugh.

My own attire had a lot to do with my first becoming aware of this. Later on, when I paid closer attention, I began to feel that I could pretty much tell where people were from by the way they walked. Italians, as is clear also from their bearing, are the most open-hearted and easy to talk to.

But Italians also have their drawbacks. I attacked their shoes. Once, when I was visiting a farm in the Italian country-side, someone asked me, "Do the Japanese wear *geta* when they work in the fields?"

"No," I replied. "We used to wear sandals made of straw or bamboo bark."

"Why straw sandals?"

"So as not to harm the earth."

"In your country," I continued, "the soil is no good and the mountains are bare of trees. The reason Italy is the poorest country in Europe is because the soil has become so hard. I kept my eyes open as I traveled from Switzerland to Austria before coming to Italy and I'll tell you, the further south you go, the fewer the trees. I noticed that very few types of trees grow in the mountains in the northern part of your country. In fact, most of the mountains there are completely bare. A mountain without a cover of trees always has depleted soil underneath. Even when it comes to farm crops, those in Italy take the most trouble to grow. The reason for all this lies at your feet. Why, you're all wearing the shoes of the Roman legions."

For some reason, the Italians, who are not very large

people, like to wear hard, hobnailed shoes—the hardest shoes in Europe for that matter. Although I had never seen the footwear worn by the Roman legionnaires, I imagined that it must have been something like this.

"Because you're trampling the soil with hard leather shoes like those used to conquer Europe, the ground has packed down so much that the soil is no longer any good. The Japanese take care of their soil. They believe that if you walk over the ground wearing hard materials like iron, you're asking for trouble, so they tread lightly, wearing straw sandals that are soft and do not harm the ground. That's why the ground in Japan is soft and fertile. As long as the ground is treated with care in this way, crops will grow naturally."

This made quite an impression on them. I was visiting an Italian commune. The folks practicing natural farming there had eagerly awaited my arrival, hoping to hear that they were doing things right. I found this a bit disconcerting, so I took the initiative by looking at the vegetation and crops and guessing at the climate. I would start out by remarking, for example, "This grass should grow better. If this is all that you're getting, then the soil is probably poor." When I broached the subject of legionnaire shoes, everyone raised their feet and said, "You mean these shoes are to blame? Show us those *geta* you're wearing."

That loosened things up and the conversation livened up after that. Once we had finished talking and it came time to leave, I was invited over to the house for a glass of wine. I told them that I don't drink alcohol, but they insisted: "Over here it's customary for people to have a drink together after a good talk. Come on and join us." The house looked run-down from the outside, but when I went inside I was surprised to find that although this was a farmhouse, it had a splendid dining room. They brought out their best old silverware and treated me royally. The ambience at the end was great. "Won't you stay with us for a week or so and teach us?" they asked.

"If you promise to come to Japan and work for me," I answered with a smile.

The Culture of Clothing ━━━━━━━━━━━━━━━━━

When I was called "Hanasaka-Jiisan" at the airport in Anchorage, I realized with surprise that that is indeed what I looked like. But then, back on the plane, as I was looking at a glossy tourist brochure of Europe, I noticed that the traditional Bulgarian peasant dress is identical to that of Japan.

Wherever I went in Europe, people reacted very favorably to me and I was often asked if what I was wearing was Japanese peasant garb. I learned that people in Europe long ago wore cruder garments than they do today.

On occasion, embarrassing as it was, I was mistaken for a martial arts master or an artist. Once, in a train station, an Italian man walked up and shook my hand. One of the young people escorting me laughed and said, "That guy looks like a gangster. Maybe he mistook you for the boss of a Japanese syndicate and came to pay his respects." The bellboys were courteous to me even when I walked over the carpets with my *geta*, and it seemed to me that the stewardesses were especially kind.

I had a reason for going to Europe in this attire.

The Roots of Clothing ━━━━━━━━━━━━━━━━━━

One thing I always feel when I work in the fields is that today's farmer in Japan has nothing proper to wear. After the war, Japanese farmers began wearing Western clothes made of nylon and other synthetics, but these are uncomfortable and hard to work in. The clothing does not breathe, making one hot and stuffy. Because this fits closely to the body, it would seem to be easy to work in, but instead it is tight and tends to give one a stiff shoulder. Perhaps this has something to do with static build-up. By contrast, the navy blue cotton clothing of the past was simple and airy. A rubber raincoat today keeps the rain out, but there is no way for the humidity to escape, so in no time at all you're drenched in sweat. This is not something that can be worn while doing heavy work. To tell the truth, I'm amazed at how farmers can tolerate such discomfort as they work. The sedge hats and straw raincoats of the past were far more appropriate. Raincoats made of hemp palm were the best of all, but today they are just too expensive to buy.

Good headgear can no longer be found either. These days, even barley straw hats are petroleum-based products, so the head practically bakes inside. And when it rains, you feel dull and heavy. For some reason, the hats worn in China are more comfortable than Japanese work hats. People around the world use many different types of headgear. Hats such as those worn by the elves of the Swiss forests may indeed be well made, even when it comes to avoiding danger. But I prefer to wear a hat made of barley straw or sedge, or a hand towel knotted as a headband in the Japanese fashion. In the winter, the best thing you can do is to wrap your cheeks with a towel under the hat. But for this, the synthetic fiber handtowels that farmers use in the West are of little use.

As for footwear, the *jika-tabi* (a heavy cloth, split-toe shoe), one of the finest inventions of the Japanese, serves the farmer well. Lately, these are even proudly worn by some American carpenters. But somehow they just do not go together with Western clothes. Also, unlike sandals, they cannot easily be slipped on and off. And when it rains, they have to be re-placed with boots, but these are too heavy and tend to leave the feet feeling hot and stuffy. Sandals made of bamboo bark or straw are still much better.

The farmer today has not a single good piece of clothing to wear. I have asked many people to come up with ideas for better wear. I've inquired as to the type of clothes people wore in the days of Japan's former glory and what the farmers of medieval Japan wore several hundred years ago, and I've even requested that tailors try making *denchi*, a sort of work vest which can be shed quickly when the weather gets warm, *monpe*, and other traditional wear. Then, just a few years ago, a new garment called a *dōi* was designed. This consists of a loose-fitting upper half similar to the Japanese *haori* and baggy work pants resembling the skirt-like Japanese *hakama*. I was sent a complimentary set. This is very neat and trim Japanese-style clothing that is good for doing farm work and can even be worn on slightly more formal occasions. Made of navy blue cotton fabric, it is extremely comfortable to wear, so I have worn it regularly ever since.

I donned this outfit and a pair of *geta* for my trip to Europe. These days, the fads in fashion seem to be emanating more from Tokyo than from the salons of Paris. Such success has gone to the heads of Japanese fashion designers, who are totally caught up in creating unusual new attire. Along with their confusion over diet, it is all too clear that the Japanese are also experiencing confusion over clothing.

I departed for Europe thus attired out of rebellion for the ways of the world and in a spirit of mischievousness, but this turned out to be a great success and won good favor wherever I went. It provided material for reflection on the culture of clothing. I ran into a tour group of Japanese women all in fine new Western outfits, even down to spectacles with attached gold chain, but nobody paid them the least bit of attention. If Japanese women were to go on trips abroad dressed in kimonos, I'm sure they would be most happily received.

When you think about it, the *haori* and *hakama*, which are surely among the most notable innovations of the Japanese, were until recently the proper dress for gentlemen and people of distinction, but they probably had their origins in the plainer clothing farmers wear, such as the *denchi* and

monpe. The farmer works in the service of God; his is an
earnest struggle with the scythe and hoe. Both in theory and
in practice, the samurai holding his sword and shooting an
arrow and the farmer swinging his scythe must be identical
in spirit and attitude. In fact, I showed people at the Zen
Center in California and farmers in Italy the correct use
of the scythe and hoe. That's when it occurred to me that
the farmer's work clothes have to be the proper dress before
God. In this sense, the *dōi*, which blends features of the
haori and *hakama*, has a clean style that braces the spirit.
I returned home to Japan confident that, as both convenient
work clothing for the farmer and proper dress, this could
well become the garb of peasants throughout the world.

The experience helped me to reflect on what the culture
of clothing is, but because man is an animal born naked,
simple diet and dress would have suited him fine. One bowl
and one robe—the traveling robe of a Zen monk even—
would do quite well. Because Japan is a country of greens
and blues, the men need wear only indigo-dyed cotton clothing from Tokushima, and the women, light blue *jindaifuku*.
Won't clothing in two colors suffice throughout the year?
These serve as both work clothes and proper dress. If they
were to spread and be adopted as standard attire in Japan,
the Japanese people would probably regain their tidy Japanese disposition.

All the Japanese ever had to do was live in houses of
wood, earth, and paper, wear the *dōi*, and drink tea. It would
seem to me that the revival of a Japanese people held in high
esteem by the world begins with these *geta* at my feet; it begins with a single work garment.

The ultimate attire for man is total, unaffected innocence.
The willingness of the Japanese to wring their own necks
with neckties is the height of folly. I was able to see this for
myself during my European voyage in *geta* and *monpe*.

The Culture of Meat and Wine

Farming for the Kings and Clergy ─────────────

One thing I noticed during the two months I was in Europe
was that wherever I took a photograph of the beautiful
countryside, a castle or church was invariably planted right
in the middle of the picture. The countryside is gorgeous
and the castles and churches fit right in, but everything is
just so beautiful that one ends up taking too many photos.

As I was wondering why, despite the great beauty everywhere, there are so few trees and only a limited variety of
grasses and herbs, it occurred to me suddenly that the reason
may well be that this land had been farmed for the church
and royalty. People drink wine at the churches, calling it
the blood of Christ. As for the kings, they wanted to eat
meat. I realized that the system of agriculture peculiar to
Europe may well have begun for the sake of the kings and
clergy. If that is the case, then an outrageous mistake was
committed. Agriculture to satisfy human desires had its
start here.

All of Europe, then, is a cattle ranch. Europe may have
shovels for digging but because it has no terraces or irrigation ponds, the soil continues to wash away, depleting the
land. It may appear as if European farmers take good care
of their land, but the land is actually being ruined. I think the
fact that this situation has continued for two or three hundred years is responsible for the impoverishment of the rural
communities. The land has been ruined by a different form
of agriculture than in America, a form of agriculture that
began with the castles and wine. When you raise cattle and
horses, you can't afford to have them get injured, so terraces

were eliminated and the land smoothed out. The slopes are green and beautiful. Europeans are proud of this, but although the cows and horses may be delighted, the land itself is crying.

When I saw this, I began to sense clearly that culture arises from agriculture. There is nothing wrong with having churches, but churches have gone wrong somewhere. In Italy, I saw carved images of Christ by the roadside which reminded me of the *Jizō** statuettes so common to the roads of Japan. These were actually cute and I much preferred them to the churches. Nothing more is necessary. The spirit of Christ wells up and causes you to reflect whenever you walk along the road or work in the fields. The image of Christ does not appear at all as you enter a church. Christ lives only within those small, sheltered images by the side of the road; he does not live within the churches.

Grapes were grown because an offering of wine had to be made to the clergymen. This too started the process of soil depletion. After all, grapes must be grown on alluvial soil—that is, soil washed down from higher land.

Thus Europe is a land of meat and wine. It is a place where an agriculture that benefits neither the farmer nor nature has been practiced, and this, I believe, is the fundamental reason underlying the impoverishment of today's farming population. It is this that has brought Christianity to a dead end.

Once I even remarked, "I strongly doubt that Christ ever said, 'make wine and drink it as if it were my blood.' " Later I learned that in Vienna, people were still talking about my rash words. Apparently some supported what I had said and others insisted that I was wrong. I was especially unpopular among those who wanted to raise cattle.

My lectures in Europe therefore revolved largely around the topics of clothing, cattle, wine, and Christianity. I would

---

* A Bodhisattva regarded as a savior of children in Japanese Buddhism.

visit farms and talk with the farmers, then in the evenings I'd go to a public hall or place of assembly somewhere and give a talk. I did this day after day. The strangest part about it all was that, although I had trouble making myself understood, I was able to talk about Western philosophy.

Yet, in another sense, all I really did during my voyage was to travel around Europe examining the vegetation wherever I went. For example, I developed a great interest in the radishes and related crucifers growing wild at the places I visited. In Japan, the plant from which the *daikon* was originally developed is the shepherd's-purse, one of the seven herbs of spring. It is no exaggeration that eating shepherd's-purse makes people more gentle and peaceful. If it is able to soften the heart, then perhaps shepherd's-purse and its sister plants can quiet the destruction of the earth. I thought it possible that this might perhaps serve at least as a starting point. This was, of course, wishful thinking on my part. In any case, I thoroughly enjoyed looking at the many different trees and grasses wherever my travels took me.

First Leg: Switzerland

From de Gaulle Airport in Paris, I flew to Zurich and arrived at a national camp there. The camp was located on a rolling stretch of green along the shore of Lake Geneva at the site of a former church. For three days I gave lectures in the morning. I had my afternoons free, so I visited the local area and also went on longer drives during which I was able to get a look at much of the pastoral Swiss countryside. The water on the lake was beautiful. Forage crops, wheat, and corn are grown in alternation on the hilly fields bordering the lake. These fields are ringed by forests. We took a road that skirts the lake, passing through a forest and emerging onto a beautiful pasture which spread out before us.

The summer wheat, which was broadcast-seeded, produced

good yields of 1,100 pounds per quarter-acre. Next to the wheat was a large field of roses where tens and hundreds of different varieties of the flower were in full bloom. A mailbox-like receptacle for placing orders stood by the the side of the road. While we were there, a young fellow came up the hill toward the farm on a motorbike, perhaps to place an order. Seeing me, he pointed proudly to his bike and laughed, saying, "Honda, Honda." Clearly, he guessed that I was Japanese. Young people must be the same everywhere.

This rose farm was the only place during my travels through Europe where I saw a farmer using a hoe in his fields. As you enter the towns in the countryside, you see beautiful castles and vineyards. But nowhere is there anyone at work. Everything is very quiet.

The only fruit trees I noticed were cherry trees. These usually stood alone and were, I imagine, for home consumption, yet it did not appear as if the cherries were being picked. On passing by, we were invited to pick a few and try them.

The Austrian countryside resembled that of Switzerland, but in addition to wheat, other crops such as oats, broad beans, and peanuts were also grown on large farms. The beans were small and used for livestock feed. However, since the fields were on unterraced slopes, growth was irregular. Perhaps half of the beans were of normal size, but the remaining half were so small as to be almost worthless. I was often asked why this should be, but it was very simple: much of the topsoil had been washed away, depleting and exhausting the land of its fertility. That's all there was to it. When I dug into the soil and explained, the farmers were soon convinced. What amazed me was that they had not noticed this before themselves.

They were overjoyed when I talked to them of sowing clover and bur clover seed mixed in with the corn and wheat seed. I saw no sign anywhere in Austria of an advanced level of farming technology, but while I learned nothing from the country's agriculture, I did learn much from the grasses I saw growing there. I found several types of leguminous plants that I thought would be interesting to use as green manure. Crossing the Alps into Italy, I saw many types of lupine growing wild. What amazed me was how completely this had suppressed other grasses. Later, when I traveled to Holland, I brought up the idea of using this lupine in that and other countries with a colder climate.

The beauty of Saltzburg and Vienna must definitely be seen by visitors to Austria. But what interested me the most wherever I went was the vegetation growing in the forests and mountains.

Natural Farming Takes Root in Italy

Summer Camp at Giannozzo's ─────────────────────

I took the train south from Vienna to Florence. As we passed
from the mountains to the plains, the scenery gradually became monotonous. The only crops I saw were corn and
wheat, and the shelterbelts scattered throughout the countryside became more infrequent the further south we went.
As the surrounding countryside became increasingly uniform,
the buildings changed from stone and wood to concrete. The
Italians riding with me deplored the fact that the country
was becoming Americanized.

I was surprised to find that there are no wickets at the
train stations. The stations aren't surrounded by fences so
one can get on or off the platform at any point. And even
though the conductor just comes through occasionally, there
are no nonpaying passengers. All the stations are spotlessly
clean and quiet. The station employees walk about leisurely,
and do not leave one with the impression that they are
working. Not a sound is made. Even the trains leave the station without a bell or whistle to announce their departure.
One locomotive that came in from a branch line at a small
mountain station was almost identical to the black engines
one used to see in Matsuyama near where I live. It appeared
to be of German turn-of-the-century vintage. I imagined that
the railroad buffs back home would really drool over this
engine. Frankly, I was impressed by such concern and care
for old things and old times.

From Florence, I headed by train for the town where
Giannozzo Pucci, publisher of the Italian edition of *The One-
Straw Revolution*, is located. A summer camp was being held

here for one week. Although he runs a publishing company, Mr. Giannozzo himself is an up-and-coming thinker. When I arrived, I found the place to be a decrepit, two hundred-year-old stone house in the middle of a combination olive orchard and vineyard. The surrounding hills and peaks were crowded with cypress trees like those in a Van Gogh painting. Here I spent a week experiencing camp life surrounded by a hundred young Italian men and women who had gathered freely from all parts of Italy. In the course of my stay, I felt as if I too was an Italian from the Roman ages and spoke straight from the heart. This was for me an enjoyable experience in communal living.

Naturally, along with the practical instruction I gave in the production of fruit trees and vegetables, I spoke about Christ, philosophy, and other topics.

Giannozzo is a singer with a beautiful tenor voice. Listening to his renditions of old Italian folk songs at the evening campfires brought up grand visions of ancient Roman gods at play in the wild.

When I think of it now, I can't help but be amazed. I spoke and drew my philosophical cartoons outdoor on large sheets of paper with a brush. A young Japanese woman by the name of Suzuki translated what I said into English. This was translated into Italian by Miriam, who had stayed at my farm in Shikoku. Some of the participants were from France, so Giannozzo translated from Italian into French. We had four languages flying back and forth, but none of this felt strange in the least. Instead, the camp was constantly rocking with laughter and gaiety. Why I can't say, but I never once felt handicapped by language.

When I was told that the cypress trees, which appear so often in Van Gogh's paintings are used for mourning the souls of the dead, I said, "In my eyes, they are lamenting the decline and ruin of the Italian mountains."

Italy was in the midst of a drought at the time. Just as I

had given Giannozzo a cotton *furoshiki** printed with an iris pattern and was explaining that this is a rain flower, a squall suddenly arose. Everyone was overjoyed. "This blessed rain has been brought by Mr. Fukuoka," they said. They repeated this so often that even I was tempted to believe it.

Although Giannozzo's place was an orchard, it consisted of little more than extensively grown grapevines among which were scattered two-hundred-year-old olive trees. Having encountered these Italians who live such a free and cheerful life, I felt both a sense of expectation that the spirit and practice of natural farming would surely take hold here in Italy and, mixed with this, a sense of envy when I thought back at how fussy and narrow-minded people are in Japan.

One day I said, "Italy has few trees. I'd like to see a virgin Italian forest." It was decided that everyone would go, so we formed a caravan of cars and drove three hours along a gently sloping highland road to a virgin forest at Pratovecchio. I fully enjoyed the beautiful countryside along the way. The gently undulating road passed by vineyards and fields of wheat. Here and there, cows and flocks of sheep were graz- ing. Every so often, there would be a single farmhouse standing alone on a ridge and children could be seen playing nearby. It was the type of quiet, pastoral scene that one sees in pictures. Nowhere did even a single billboard or telephone pole mar the view. As for coffee shops, which seem to be all over the place in Japan these days, we were on the road for several hours and we stopped at perhaps the first or second place we happened upon. I'll never forget the Italian coffee I drank below the gigantic apple tree there.

Even in Italy, which is said to be the most Americanized of the European countries, as soon as one gets out of the cities, the beauty and serenity of the countryside that has long since vanished from Japan remains exactly as it was in

---

* A square cloth used in Japan for wrapping and carrying articles.

the past. Country roads are not paved anywhere. These are bumpy and winding, and make, I felt, for a more interesting and enjoyable drive.

While thoroughly impressed with the preservation here of Italy's past, I searched for the basic cause behind the stagnation of European agriculture. I suspected this to be depletion of the soil. While wondering thus what plants would be best for reviving these lands that were approaching a desert state, I was invited to speak at the Agricultural Academy of Florence. When I went over, I learned that the president of the academy was trying to introduce tropical plants. I was given a tour of the botanical gardens, where many rare and unusual plants were growing. Later, after prefacing my remarks at the lecture with "let me be quite frank with you," I asked whether it wasn't far more important to dedicate one's full efforts to preventing desertification by enhancing soil fertility and to bringing back the plants native to Italy. I laid out my views, saying that, if it were up to me, I would do such and such. The dean of agriculture was overjoyed and immediately offered to translate *The Natural Way of Farming* into Italian. This boosted my self-confidence.

The decline of agriculture is a consequence of the impoverishment of the earth for which the farmers themselves are culpable. At the same time, it is also a tragedy in which farming has become embroiled in politics and economics. This problem is the same today in any country, but I found a particularly good example one day while near Venice when I saw conflicting reports in two different newspapers—one published in Austria and the other in Italy.

The Viennese paper reported that fruit in that city was so high-priced because Italian farmers were limiting production. But earlier, when I had crossed over into Italy, I had noticed heaps of ripe cherries and pears picked and left to rot in the orchards. The farmers there said that they had been told to dump their fruit because prices in the cities were too low.

The middlemen, those at the core of the distribution system, turn to the consumers in the cities and tell them that produce is expensive because the farmers don't grow enough. At the same time, they tell the farmers that fruit isn't selling well in the towns and instruct them to produce and ship only good fruit.

It is all very simple. The consumers and farmers are made to dance to the tune of the news releases put out by the merchants who exist solely to buy small quantities of high-grade products cheaply and sell them dearly. As a result, people in the cities eat expensive fruit while farmers sell their produce at low prices. Later, while I was still in Europe, I heard the news of how a truck carrying Italian wine into France was attacked by French farmers. Here we have the tragedy of farmers stirred to violence by misinformation. The price of farm products everywhere is under the control of the merchandisers and the media.

Milan Rice

After the camp near Florence, I toured the rice-growing district near Milan. Rice sold under the Ivo Totti label is famous as natural rice not only in the local Milan region, but in other European countries as well. The 300-acre farm was the location of the movie *Bitter Rice*, starring Sophia Loren. The owner of the farm is getting on in years but is still in good health. He has a deep interest in natural farming, and was a constant and invaluable benefactor throughout my entire stay in Europe.

He offered a Japanese youth who had been moved by *Bitter Rice* to meet the beautiful, young actresses who had appeared in the movie. We had a good laugh together when we saw that we were now all old men and women.

He was excited about sending a current through the world of European agriculture with the natural rice grown in his fields, so I intend to pay close attention to his future efforts.

Rice in the region about Milan is no longer transplanted as in the past. It is grown today by direct seeding in flooded fields. Because the fields are constantly kept under a deep cover of water, I suspect that root rot must be severe and harvesting quite difficult.

Next I visited what may well be Italy's largest farm, consisting of some 50,000 acres of land, of which about 750 acres are farmed. My book (*The One-Straw Revolution*) had been read here, and part of the land converted to the cultivation of summer wheat in a clover cover, all with fairly good results. Seeing that such a vast stretch of land was being left uncultivated, I suggested that upland rice be grown on the irrigable fields. The owner of the farm, a real go-getter of a woman, immediately asked me to send over a hundred kilos of upland rice seed and even offered to have one of her three sons accompany me to Japan. As she showed me around the farm, she gave out a steady stream of orders to her sons and the farm hands.

Many Italian farms seem to run about 25 acres or so in size. That scale is more stable, and in fact there seemed to be few successful large farms. For example, the neighboring 300-acre farm employed three farm hands to grow corn and wheat, but was in desperate financial straits. Later, when I had a chance to talk with the three farm hands, they told me, "The situation is so bad that we couldn't possibly ask for a raise. But we have vegetable gardens for our own families off in a corner of the farm where we practice natural farming, and that gives us something to live for." They asked me all sorts of questions right there in the fields.

Although Italian agriculture is clearly on the decline, I found examples everywhere of people who had begun practicing natural farming, led by a core of farmers who want to inject new ideas and approaches to farming, by individuals aiming for a natural diet, and by those trying to restore nature.

I was invited to a place called Renate. Some folks here who ran a Zen center wanted to start up a natural farm and sought my advice. As in America, the soil on the upper slopes of the hills was eroded and poor, so I talked about green manure trees and sod cultivation.

I also visited the farm of the Allegtori group, which became widely known for its fight against conscription during the Second World War. One thing I found interesting there was that fruit trees of the four seasons are planted radially about the house. The focus of our discussions was the rotation of vegetables. I always stopped and spent two or three days wherever I went, so our conversation often got quite technical and involved.

On to Austria

From Milan, I then traveled toward Venice, where I spent several days touring the local area and getting an idea of the current state of Italian agriculture while staying at the Moro Farm. After that, I went south a bit, then crossed the Alps again by car, arriving at Innsbruck in Austria from western Italy. This was to participate in a summer camp organized by Michio Kushi of Boston. During the entire trip, I was so captivated by the beauty of the alpine ravines that I was unable to utter a word or take my eyes off the passing scenery for even a moment.

Mile after mile, I saw vineyards and apple orchards at the bottom of these mountain gorges. Steep rocky crags soared up on all sides, but the fascinating variety of peaks and rock promontories was totally impossible to capture on film. When I had visited the virgin forest at Pratovecchio in the center of Italy, the park ranger there had told me that the authorities were uncertain what to do about Italy's rocky mountains. I at once suggested the use of Japanese creepers, but when I got a chance to see these towering limestone mountains for myself, I was taken aback by their beauty and sensed that this was destruction* on a scale about which man was powerless to do much.

"Even so," I thought, "It won't do to merely sit in wonder doing nothing. It won't be easy but there must be something that can be done. One idea might be to take a helicopter up there and scatter the seeds of creeping vines and other plants that grow well on rocks. Once shrubs have begun to take, the area might be restored to a natural state after several hundred years." These and other thoughts ran through my mind as I enjoyed the beauty of the gorges.

Not only are Italy's mountains lacking forest cover, the country's flat areas have fewer trees and shelterbelts than Austria. This is an indication of how poor the soil is. I viewed that as the main reason behind the sad state of Italian agriculture.

Although most of the fruit trees I noticed growing along the bottoms of the steep alpine valleys were still young trees, they were being grown as single stem plants such as I advocate. These had an almost natural form which had my full approval. I had heard that fruit tree cultivation in Europe was generally most advanced along the Mediterranean coast, but I found that techniques there were not more advanced than in Japan. In fact, growers had a more easy-going approach, so if it were suggested that the trees be grown in a natural form rather than in the existing form, they were able to make the switch rather easily, both mentally and technically.

Later in my trip, when I encountered the several hundred enormous apple and pear trees at the Nelissen Farm in Holland, I gave some advice on how these could be corrected to a natural form. The trees had been left alone and had not fallen prey to bad pruning, so it seemed likely that they could easily be brought around to an excellent natural form within two or three years.

---

* These mountains had been mined since the Roman ages.

The grape, the number one fruit throughout Europe, is grown on wire trellises as in Japan, but the vines are not heavily pruned and sometimes they are even trained on one or two stems without using trellises at all. When I saw these, I began to wonder how they could be approached to a natural form.

It was toward evening when we descended from the pass and at last entered the city of Innsbruck. This town, site of the 1976 Winter Olympics, is a beautiful place surrounded by cloud-enshrouded mountains. A large public hall had been rented as the site of the summer camp for a full ten days. The camp was a big success, with about one thousand people attending from all over the world.

Lecturing in Vienna

Let me return a bit to my visit in Vienna, where I stayed before going on to Italy. A little incident occurred during a talk I gave there. The lecture had been hastily arranged and it was thought that only about three hundred people would show up. But as we were beginning, people continued to arrive and it became clear that not everyone would be able to get in, so the lecture was postponed for thirty minutes and the location moved to a larger hall. It turned out to be quite an emotionally charged meeting.

Ten to twenty minutes after I'd begun talking, one young man stood up. "I came here to learn about natural farming," he said. "But all you're talking about is Western philosophy. I'm not here to listen to you speak of philosophy." This is the same kind of thing that young people tell me in Japan.

I had started talking about philosophy because, when I arrived in Vienna and saw all the churches there and when I looked out at the audience and saw that they all had the faces of musicians, it just seemed to be a good way to start off my lecture. I saw almost no one in the audience who looked like a farmer; they all looked like townspeople to me. That's why I began by saying: "Austria is beautiful but a land of spurious green. This is a country of cattle and grapes. The agriculture here arose to produce meat and wine. This is not farming for the natural earth; it is an agriculture for the royalty and clergy. That is why the earth is poor and barren today. If agriculture takes a wrong turn, then culture also goes awry. This mistake began with Descartes. Such destruction is the penalty for the crime of thinking that nature exists because man exists and of sacrificing nature for man."

Then, in the middle of this, someone stands up and hands me a wallop: "I came to hear you talk about natural farming, not philosophy." This infuriated me, but it also spurred me on. I launched into a harangue.

"You may say that but do you intend right now to practice natural farming, and are you able to? Speaking from thirty years of personal experience in Japan, not even a single farmer in my immediate neighborhood practices natural farming. Do you know why? There's a reason for this. Even if you yourself intend to go into natural farming, do you think consumers in the towns and cities will buy crooked eggplants and vegetables full of insects? If the people of Vienna don't understand, you won't be able to support yourself. One individual may think of practicing natural farming, but it's not something you can do right away in a field. In order to change the farming practices of a single farmer, the entire social fabric must first change. Natural farming is not simply a question of agriculture. It is a problem that concerns politics, economics, and people's ways of thinking and living. It concerns everyone—consumers in the cities and farmers alike. That's why to reform one thing, everything has to be changed. Did the chicken come first or the egg? Well, the key to changing everything at once lies in philosophy. If one thing changes, everything changes. Unless all things change, nothing changes. If the philosophy of all the people of the world doesn't change, if the thinking of the people of Vienna doesn't change, then no one will be able to practice natural farming. Unless all the problems are solved, not even one thing can be done. The methods described in *The One-Straw Revolution* can resolve all the agricultural problems, but unless reforms occur in all areas—Western philosophy, thought, and religion—even so simple a thing as this cannot be done. No one will be able or willing to practice even such an easy method of farming as this."

This received a thunderous ovation. The applause just wouldn't die down. There were cries of "Encore, Encore." I loosened up after that and did get around to speaking about natural farming, but once the lecture was over—it must have been ten or eleven o'clock in the evening—the audience didn't budge to go home. The doorkeeper at the hall tried to push everyone out, telling them, "It's time to close the doors. Please go on home." But people just wouldn't leave. We had a second meeting of sorts outside in the lobby. Then, when we finally left the building, everyone went over to a kind of dining hall where we held a third session. We went on like this until past midnight.

"Gosh, my talk really went over well," I thought. It was a great feeling. But later it occurred to me that this was Vienna, the music capital of Europe. People here had the habit of encoring at concerts and recitals. It was even possible that all that clapping may have been done partly out of politeness. Still, the audience had taken to what I had said.

The interpreting done there was very good. A Japanese interpreter from the embassy in Vienna came over to translate for me. I realized there, and also later in Salzburg and Innsbruck as well, that a great deal depends on the interpreter. The interpreter at Vienna had read *The One-Straw Revolution* and, during the lecture, he always translated exactly what I said, even when I was groping for words. He refrained from inserting any of his own thoughts and did an excellent job of relaying my message to the audience.

At Innsbruck, an attractive young French woman began

interpreting for me, but she appeared to have trouble understanding the more difficult points and was joined by a second interpreter. The Innsbruck public hall, which is where the opening ceremonies for the Olympics were held and is also often the site of international conferences, was equipped to handle simultaneous interpreting in five languages. In addition, the speaker's platform had a device that allows you to make a drawing on a sheet of clear plastic at your desk and have this immediately enlarged and projected on a large screen behind. I used this to get what I was saying across with my particular brand of cartooning. I usually manage to make myself understood when talk turns to religion and philosophy by depicting familiar things with these sketches.

The Old Man and the Mill

North to Holland

From Austria, we followed the Rhine through West Germany. I had thought the river would be clean and beautiful, so you can imagine my surprise when I found the waters muddy. This is evidence that the mountain soil is continuously being washed away. Vineyards are located only on soil deposits at the foot of mountains and on mountain plateaus. There seemed little future in these famous vineyards along the banks of the Rhine. A mistake must surely have been made in the means used for protecting the land.

In the West German lowlands, a region of rolling hills planted in wheat and corn continues almost without end. Much of the land is forested—primarily woods of red pine with small bushes and trees. This means that the soil is not very fertile, which explained the abnormal early ripening of wheat I noticed.

Upon crossing into Holland, the land becomes completely flat. This is almost all pastureland. Houses are scattered here and there throughout the countryside. I found it hard to believe that this is the world's most densely populated country. To me, it appeared as a spacious, rural land. Shelterbelts of poplars are more common here than woods, but I suppose this has a lot to do with the national character.

Although wherever you go in Europe, cutting a neighbor's tree—even if it blocks the sun—can land one in court, just shortly before I arrived in Holland, someone had taken their neighbor to court for not allowing him to cut a single tree, and this flared up into a big dispute carried in the papers. In Holland, where the sun's rays are weak, this was a personal cry for sunlight. This incident underscored the differences between Europe, where cutting a single tree can develop into a major controversy, and Japan. The sentiment that seeks after and cherishes sunlight in this way must be what built this country of tulips—flowers that bloom only after receiving the full light of the sun.

While in Holland, I was taken to see a national park with huge trees. I was amazed that such a forest should remain standing in Holland, but I soon learned why. A large number of farmer's cottages are carefully preserved at the center of the forest as a historical exhibit. The homes of the Dutch long ago were amazingly small. The bedrooms, for example, measured hardly more than a couple of yards square, a clear indication that people back then were of small stature. It certainly seems plausible that the physical proportions of the Dutch grew when they began raising cows and making butter and cheese.

Naturally, the park had a few windmills and also a number of Dutch-style wooden buildings. The Dutch were a proud people who cherished sabots and houses made of wood.

My impression, after receiving a tour of central and northern Holland, was that this country has a cattle-oriented agriculture. From what I had seen, I concluded that as one moves north from Italy to Austria, West Germany, and Holland, the variety of cattle does change somewhat, but even more importantly, the number of cows grazing increases. Herds of from several dozen to several hundred head become apparent. I was troubled, however, by the fact that the color of the countryside seemed to grow a paler green in areas where there were larger numbers of cows. When I later asked one farmer about this, he told me that as the number of head being raised increases, dairy farms have become dependent on concentrated feeds such as corn and wheat, making operations financially tighter.

"I suspect that the basic cause for the decline in productivity is depletion of the soil," I said.

"That could well be," he answered slowly, as if turning this over in his mind.

Small dairy farms with twenty or thirty head of cattle were more stable; it was the farms raising hundreds of animals that were seeing hard times. In America too, very few ranches have thousands of head; the great majority of dairy and beef cattle operations have at most several dozen head.

These are difficult times for livestock and dairy farming in Europe today. And the larger the farm, the more difficult the situation. In northern Holland, the average 100-acre crop farm rotates wheat, oats, potatoes, broad beans, and rapeseed (used medicinally in Germany), but one farm in two is being faced with the harsh reality of having to go out of business. One highly competent farmer I talked to who uses large machinery to work 100 acres with his son lamented that farming a generation ago with maybe six or seven cows and horses in which all the members of the family helped out was better. It was very helpful learning why he felt such a strong interest in natural farming. I heard many things in Europe that would prick the consciences of proponents of modern, large-scale agriculture in Japan.

The Nelissen Farm

It occurred to me then that the difficulty banks were experiencing in recovering loans made to farmers may have had something to do with their willingness to lend Thomas Nelissen the equivalent of $400,000 to try his hand at natural farming. Thomas was a young Dutchman who had stayed at my farm for three years a while back. Upon returning to Holland, he spent a year traveling around the country and teaching people how to set up home vegetable gardens. This worked out well and became quite popular, which helped convince the banks to loan him the funds he needed to set up a natural farm.

I had once told him that the pond in a Japanese garden should be dug in the shape of the Japanese character for heart (心). I don't recall whether I heard this somewhere myself or stumbled upon it on my own, but that is how I built my garden at home. If you dig a garden in this shape, then the pond is wider at certain points, leaving some areas floating free like islands. What you have then is water flowing downstream, a pool, a sea, and islands. I told Thomas, "If one patterns it after the character for heart, then even a novice can make a pond." Thomas returned to Holland and traveled about the country, instructing people to take up a spade and dig up their lawns in the shape of hearts. In this way, high and low ground is created, so you have mountains, rivers, and valleys. When water is made to flow from the left side of the "heart," this immediately gives a Japanese pond. A garden can be created in this way without requiring the services of a gardener.

One should then scatter over the garden a mixture of the seeds of many different vegetables: *daikon*, Chinese cabbage, burdock, carrot, and so forth. The garden has its hills, plains, and shores. Water dropwort, shepherd's-purse, and hornwort seeds that fall near the water's edge take well. Vegetables such as hardy beans and *daikon*, carrots and squash grow in higher, drier places, while cucumbers take in the slightly moist areas.

You just let the seeds decide for themselves at what height and in what degree of ground moisture to grow. At first you will need a lot of seeds, which must be mixed together and scattered throughout the garden. Once this has been done, you will notice a mix of plants emerging and growing. The Western idea of a garden is to plant grass, create an artificial stretch of green, and enjoy this. However, the typical pocket-sized garden in Japan contains a bit of everything. It may be small and appear the very picture of confusion, but this very confusion is a source of great fascination. In this way, one may have a productive garden going within a year's time.

The reason this type of garden has become even more popular abroad than in Japan is because of the dearth of vegetables in Europe and America. The notion of the division of labor has become so entrenched there that few people have home gardens. Instead, everyone buys and eats produce transported from far away. In America, for example, California supplies fruit and tomatoes to the entire nation. This is the way things are set up. These are not fresh or tasty, and because they are uniform market products, the variety is very limited. This is true as well of vegetables. No wonder then that the woman of the house never had any real interest in growing vegetables. But the backyard vegetable garden is something one sees morning and evening. And here, one can grow a variety of vegetables.

I saw several of the vegetable gardens created in Holland under Thomas' tutelage. One of them had been set up at the home of a priest. When I visited him with Thomas, the priest told me about himself: "Up until a few years ago, I had been active in the church and even played the pipe organ. But then I became neurotic and felt as if there was no longer anything to live for. That's when Thomas came by. He taught me how to set up a vegetable garden. While I was working up a sweat spading the dirt with my wife and growing vegetables, I regained a joy for life. From the vegetables I learned what it means to be alive and felt the joy of being active. Lately, I've come to enjoy life. In church, too, I'm able to preach now with renewed vigor."

He seemed overjoyed that Thomas and I had come to see him and treated us to lunch. He even played the piano for us, although I know nothing about music. We ate lunch while listening to music, then afterward he showed us around his church. It was a large, concrete, American-style church.

"I don't care much for this type of building myself," he said. "But the trustees told me to build it this way, so I did."

He told us to sit down and listen because he was going to play the pipe organ for us. There were five or six of us.

We all sat down at the center of the church, which reminded
me of a large temple, and listened to him play the pipe organ.

When he was through, he came over and said, "Just tell
me what you felt."

"At first, I had this feeling of fear, as if I were being men-
aced and intimidated. But later on, at the end, it felt as if
I was listening to a lament of God. I didn't find it in the least
enjoyable or entertaining."

He thought this funny and let out a hearty laugh. Not in
the least offended by what I had said, he then offered to take
us out to see a beach at the seashore. We got into his car
and drove over to a beach crowded with people, where we
walked around for a while.

He had a specific reason for bringing us to the shore. Much
of Holland is below sea level, so taking good care of its seashore is vital to the country's well-being. This is why it works
so hard to protect its breakwaters. The priest had wanted to
show us how the country is doing this. He took us to a test
station where vegetation is planted and research conducted
on different varieties of plants, and asked for my opinion.

I saw where the Dutch government had planted and was
growing various types of saplings, but what caught my attention were the unfamiliar vines and crucifers growing wild on
the sandy beach. "It would be interesting to plant these,
wouldn't it?" I said. Although the scientists here were obviously working hard at raising and transplanting saplings,
the plants used to hold the sand did not appear to me to be
doing all that well. The vines and shrubs growing wild in
between looked healthier. It seemed to me that these would
probably do quite well.

Thomas now operates a 75-acre natural farm. The name
of the farm is "Yakuso." This is how "hyakusho," the Japanese word for farmer, sounds in Dutch. He lives in a house
like a castle. One third of the farm consists of enormous
apple and pear trees, another third of wheat and naked barley

fields, and the remaining third is planted with a great variety
of vegetables brought back from Japan. He has succeeded
in growing the wheat and barley in a cover of clover.

The apple trees were abandoned trees measuring perhaps
fifteen feet high, but because they had not been improperly
pruned it seemed as if they could be restored to a natural
form without great difficulty. I showed Thomas how this
could be accomplished in two to three years.

Vegetables such as *daikon*, Chinese cabbage, and sweet
potato were doing very well, while the setting of fruit on the
squash and cucumber left just a little to be desired. Thomas
was concerned because the *daikon* he grew was of a type
called Tokinashi, which he felt might be too small for the
market. Also, since most people in Holland don't know how
to eat *daikon*, along with selling his produce, he had to show
customers how to cook it, all of which seemed like a lot of
trouble to me. He would even demonstrate how to prepare
grated *daikon*. He had over a dozen hardworking young help-
ers and so, all in all, I felt that the prospects for the future
of the farm were good.

Japanese bedding (*futon*) was selling well in Europe, as in
America, so three young women on the farm were busy making *futons*. Thomas remarked with a laugh that daily expenses
were covered by baking bread. Natural bread made by
grinding naturally grown wheat from the farm at a windmill
was proving a commercial success. I've heard that lately they
are even shipping this out to West Germany.

Thomas had also assembled the equipment needed to make
miso. "Someday we're also going to make miso and tofu
here," he told me enthusiastically. He had rebuilt a large
barn on the farm, fashioning the second floor into sleeping
quarters for the young people who come here to learn natural
farming and making the first floor into a meeting hall.

For three days, I gave a series of morning lectures at this
hall to two hundred people from all over. In the afternoons I
offered some practical instruction. When the camp came to

an end, Thomas made a closing address in which he called
me the father of European nature. This he followed by such
touching words of parting that even I was moved to tears.

I then gave the young people assembled a final word of
encouragement: "Even without me, you have Mother Earth.
In Japanese, I understand the word 'hyakusho' (百姓＝farm-
er) to mean "she (女) who gives birth (生) to all things (百)"
(the Earth＝the Virgin Mary). If you obey the will of the
Virgin Mary who gives birth to all things, then even without
me, you will without fail become a center of natural farming
in Europe."

The Old Man and the Windmill ————————————

Just like everything else, the pretty windmills of Holland are
getting swallowed up by the wave of modernization. Thomas
took me over to visit the windmill where he has his wheat
milled. When I saw the sturdy, powerful figure of the old
man tending the mill, I felt somehow as if I understood the
source of his vitality.

When the blades of the windmill, each of which measured
dozens of yards in length, started turning and generating
terrific eddies, the old, four-story, thin-walled brick structure
began rocking back and forth so hard that even now I can
recall my fear that it would topple over. A large post passed
through the center of that narrow tower, running all the way
from the cellar to the roof. The top of the post was attached
to several enormous wooden gearwheels measuring perhaps
two or three yards in diameter. These meshed when the blades
of the mill started to turn, creaking loudly and giving off
smoke. The smoke apparently makes it unnecessary to oil
the gearwheels. In the cellar, a huge stone mortar attached to
the post turns, tossing up flour. The miller has to constantly
get out onto the second-floor landing to change the orienta-
tion of the mill depending on the direction of the wind, and

he must adjust the rotational speed of the gearwheels in accordance with the wind strength. From what I was told, these adjustments determine the quality of the flour made. Operation of a windmill apparently requires the same arm strength and mettle as the captain of a large sailing ship. The old man told us of a youngster once who had climbed onto one of the long windmill blades and was killed when he was blown off by the wind while spreading the sailcloth over an auxiliary blade.

From what I was told, becoming a top-rate miller requires the training and experience of three successive generations. This old man who had weathered such adversity had a robust face overflowing with confidence that called to mind the title "master of the windmill."

"The Dutch windmill will probably soon vanish," he said. "There just aren't any strong oaks like this one left anymore. Even if there were, no craftsman today could ever get just the right angles on these gearwheels. And it would be next to impossible to train young people with what it takes to run a windmill."

"Worst of all," he lamented, "today there are efficient flour mills made of iron so windmills are no longer needed." He paused, then added, "But you can't make real flour except with a windmill." Raising his head, he declared, with a look of triumph, "People are fools to ignore all that's good about such wonderful windmills as this."

He told me also that true flour can't be made except in stone mortars because these do not generate heat. People think they have improved the process of milling flour. They believe that they have scientifically developed means for easily producing good flour. We are already in an age in which computerized mills grind our flour, and the pure white product is packed in tanks or huge transport vessels for shipment. But that flour is not real flour. Just what do we think we've accomplished? Hasn't all we've done been to lose sight of real flour and real people?

Dedicated totally to milling flour, the master of that windmill was a miller, and yet he was more than just a miller.

Praying for the safety of this venerable old miller, I made a drawing on the large post of him looking down at people in a cave below. He was delighted. When we left, he gave us a parting wave of his big hand and, with the rays of the setting sun full on his back, said, in a clear, resonant voice, "Let's both do what we can."

This made me reflect once again on what it is that we must not lose.

European Civilization at a Standstill

The European Culture of Food ————————————

I wanted to find out what Western cooking is all about, but during my tour of the European countryside, I found it totally incomprehensible. I was interested in learning what farming families out in the country normally eat. They eat—oh, I don't know—some kind of soup or stew made of vegetables and a bit of grain floating in milk and wheat flour. To me, it looked very unappetizing. After a meal of this, I didn't feel as if I'd really eaten. People eat or don't eat breakfast as they feel like, and the woman of the house doesn't rush about busily in the kitchen before lunch preparing the meal as do Japanese wives. It seems as if all they do is slip into the kitchen just before mealtime and bring out one or two plates or pans. I don't know whether it's because everything is self-service at home, but people are so calm and sedate that meal-time is quite dull and leaves something to be desired. To expect a delicious meal is to invite disgrace upon oneself. You serve yourself one or two different dishes, generally something resembling a stew, sprinkle some condiments on top and mix. What you are doing, essentially, is preparing the food in your own plate. Then, once you are finished eating, you wipe the plate clean with bread and wash the dishes in the kitchen. The whole process is just too dreary.

Of course, all I did was to make a quick tour of the European countryside. I didn't sample the refined cooking of the cities, so I'm not really in a position to say anything. But I did get an impression of the Western diet as being some sort of artificial food prepared based on certain principles of human nutrition. Nowhere did I come in contact with local cooking that brought out the flavors of nature. The best one could expect is boiled fresh-water fish. Why there is not the same careful attention to nature and man as in Japanese cooking I cannot say. Perhaps this is only natural. In the West, man and the body come first. Food is for the flesh. As long as something is nourishing, that is all that counts. It is not the plates and cutlery that one eats, after all.

Still, at the farmhouses I visited, my hosts proudly showed me dinner table pieces that had been handed down to them by their ancestors. These are kept safely in display cabinets, however, and not used in everyday life. Such dishes are valuable objects for man, but not receptacles for meals. Here nature, the food, the plate, and man are all separate, unlike in Japan where they are one. To begin with, the selection of food is limited, especially the vegetables.

Do you know why everything is self-service? Because everywhere, it's "ladies first." The wife doesn't cook, so everyone prepares his food to suit his own taste. Women take no interest in cooking on account of the poor variety of materials. Isn't it because men and women eat separately that meals have become totally self-service?

Europe and America have too small a variety of vegetables, and the quality is poor. There seemed to be far too few root vegetables in particular. Burdock grows wild in the fields but no one realizes that this is a food. There are no taros. The yams are sticky and not fit for human consumption. People view devil's tongue and "barbarian" foods that grow in mud such as arrowhead and lotus root as outlandish. There is no bamboo, so naturally no one knows of bamboo shoots. There are few tasty sweet potatoes. The sweet potatoes that are grown are hard and tasteless—hardly what you'd expect of a sweet potato.

Westerners thus consume very few root vegetables, which as a group are essential to proper physical growth. Perhaps this is one of the reasons why they differ so from Orientals. The dearth of roots in the diet may have something to do with the inability of people in the West to see the depth of nature's roots.

Of course, I myself am quite careless about diet, so I really have no right to criticize others about it. What seems strange to a Japanese farmer, though, is that Westerners place man first and foremost with regard to everything—food, clothing, and shelter. At first glance, Westerners appear to treat nature with care, keeping it clean and beautiful. But this is only superficial. A closer look will show that while this attention may demonstrate a care and concern for man himself, there is no genuine concern or spirit of harmony directed at true nature itself.

Isn't it the same with the shape of any plate, cup, or spoon that you look at? Western plates are large so that you can mix all the ingredients on one plate, which makes for efficient eating. Cups have a handle, making them easy to hold. Spoons are conveniently designed for carrying food to the mouth. In Japan, however, individual dishes are served in an assortment of plates and bowls. One serves a small amount of choicely prepared food in a small plate and savors it as one eats. Even the size of Japanese teacups differs with the type of tea. Teacups have no handles, so one does not boorishly hold the cup in one hand, but balances it in both hands, reverently savoring the taste before beginning the meal. With a pair of chopsticks, one is able to dexterously pick up a single grain of rice, bring it to the mouth and savor it. You do not rush through the meal. True, the ecological damage caused by the excessive use of disposable chopsticks is all too clear, but whereas chopsticks will again become trees if returned to the soil, the petroleum and mineral resources spent on the fabrication of metal products cannot be recovered. The delicate method of savoring Japanese cooking represented by a pair of bamboo or lacquered chop-sticks may itself be regarded as the soul and essence of the culture of diet. The Japanese diet shall surely perish the moment the chopstick is abandoned for forks and spoons.

Japan's farmers are people who exist by the grace of

nature, so it is only proper that they take good care of nature, their creator, and treat its soul with reverence. Westerners, however, see man as surviving by the conquest of nature, so they think it only fitting that nature be rebuilt to suit his convenience. The drive in the West to protect the environment is also for the benefit of man, not nature. The discrepancies between East and West in food, clothing, and shelter can all be traced back to this fundamental difference in view.

Why is it that in the West, the castles, cathedrals, and even the homes are all made of stone or brick? Well, I suppose that the direct reason may be erosion of the land and exposure of the underlying rock, making stone readily available as a building material. But I personally think that an even more compelling reason was the notion of the survival of the fittest, of the strong prevailing over the weak, which motivated people to protect themselves. A defensive attitude translates directly into an aggressive mentality. Houses made of stone or brick are dark and cold.

I even felt as if the gloominess of West European civilization arises from these prison-like stone castles and churches. A great deal of firewood is needed to make bricks. Because the Great Wall of China and the ancient cities in the Middle East and along the Silk Road were all made of bricks, trees disappeared from these areas and the soil died. This destruction of nature brought about the decline of human civilization.

Signs are in evidence everywhere of how nature was sacrificed for the construction of Europe's cities. At Salzburg, for example, one can see large trees growing outside of the castle, but there is no dark soil at the base of the trees. If one looks for it, there is clear and unmistakable evidence that this soil was once ruined. The crisis of stagnation associated with this culture of food, clothing, and shelter clearly comes from the decline of nature.

Flowers and Beauty ————————————————

Still, I found Europe outwardly a place of great beauty and cleanliness. Woods are common in the cities as well as the countryside. Roses blossomed wildly in the flower gardens on country farms, while flower boxes in urban homes were decked out with beautiful live flowers. Wherever I went, it felt as if the whole town was filled with flowers.

The townsfolk brought rattan chairs out onto the stone pavement and there leisurely drank tea. Things were so quiet and tranquil that I wondered where all the working people had gone. It seemed as if time had stopped. The sensation was such that even I forgot that I was here a stranger in a strange land.

While traveling around the countryside in Europe, one thing that caught my attention was the striking beauty of the wild flowers. The flowers by the roadside in Switzerland and Austria in particular had such splendidly large petals that they looked more like cultivated flowers than wild flowers. I suspected that these were either the native stock of beautiful flowers or even the wild variants of artificially bred flowers.

Lovely meadows stretched out like fields of alpine flowers. But once I had gotten accustomed to the sight, something seemed to be lacking. I had a hunch that this might be con-nected with the fact that I saw few plants which resembled wild grasses. In California I had come to the conclusion that undesirable pasture grasses had driven off the native grasses and become one cause for the desertification of the land. But in Europe, perhaps the cultivation of flowers had wiped out the native grasses and resulted in a simplified vegetative cover. Of course, this was little more than just speculation on my part, but the presence of so few wild grasses seemed to indicate that something was amiss. Or maybe I was simply unaccustomed to the sight of such a riot of gorgeous wild flowers in full bloom.

While on the subject of flowers, I remember all those weekend vegetable gardens. On the outskirts of towns and cities, one often sees tens and hundreds of miniature gardens no more than a few square yards in size with cute little huts on each plot. Although vegetables are also grown in these gardens, it seems as if flowers and spice blossoms are more common. I imagine that people pass pleasant Sundays here. The enchanting sight of these lilliputian hideaways brings to mind the life of the common townspeople. The intensity of this affection all Europeans have for flowers and the loving care they lavish on them is something very special.

In the summers, everyone takes to the mountains, forests, or sea to enjoy nature. But their way of enjoying nature differs from that of people in the Orient. Rather than nature, what they really seem to be enjoying is people in a natural setting. The Japanese carouse in self-oblivion under the cherry blossoms in spring, but Westerners don't step out of themselves and blend freely with nature. They are always in full possession of the self.

I pointed out earlier that the nature in the West is beautiful and quiet. That quiet differs somehow from the tranquility of Japan. It does not give rise to the sad, lonely Japanese ambience of *wabi-sabi*. I wonder why. Could it be that the spirit and mystery of Mother Nature that frees us from our sense of self does not exist in the forests and flower gardens of the West created by egoistic man?

The Japanese too have a strong affection for flowers and have even raised *ikebana*, the practice of flower arrangement, to an art of which it can be justly proud. But even *ikebana* has changed. Lately, steel wire is used and dried branches are colored with paint, for example. People seem to be calling this an art of self-expression these days, but to my way of thinking, the true art of *ikebana* has been discarded and replaced by an artificial, Westernized sense of beauty.

A while back, someone came over from one of the main schools of *ikebana* for an interview. I told him that I was entirely dissatisfied with the state of *ikebana* today: "Don't you think that the soul of nature was better portrayed in the days back when one spoke of the *ten, chi,* and *jin* (heaven, earth, and man)?* Even that too can be seen as just a shallow understanding of nature. What would happen if one were to see nature from directly above, behind, or below? For example, if one looks from above, the leaves on all plants emerge in a spiral. Wouldn't it be interesting to think of a counterclockwise *ten, chi,* and *jin*?"

I spouted off right and left throughout the interview, but when he sent me a copy of the issue with the write-up on me, my critical comments had been edited out. All the article does is to show the scenery on my natural farm.

What I am worried about is that man no longer sees what beauty is, from where it arises, and why something is beautiful. Decorating a room with lovely flowers or paintings is not itself what beauty is all about. It is not that there are no beautiful flowers outside, but rather that we no longer are able to see the beauty of nature's flowers. People must reflect on this forfeiture of the self that compels one to arrange and decorate flowers. Instead of growing and enjoying flowers created by man, instead of displaying flowers in a room or learning how to arrange them, I think that people should first grieve over the loss of beauty in their own hearts.

The flowers in the teahouse** are not beautiful because they have been handsomely arranged. If a single flower is brought into a simple teahouse free of any adornment, then even a wild flower will immediately appear to be a gorgeous *chabana* ("tea flower"). All of which means that when people look at flowers, they generally do not see them. Beauty does not come from the flower itself. This is just the beauty of the cherry blossoms seen from the jailhouse window.

---

* Traditional names for the three main branches in a form of *ikebana* known as *shōka* or *seika* ("living flower arrangement").

** I am referring here to the small, enclosed structure in which the tea ceremony is held.

What we have here is a humanity no longer moved by true beauty smugly saying that a beauty worthy of man has been created in cultivated and artificial flowers. The stagnation of European culture arises, I believe, from the fact that no matter how much one decorates the gardens and windows, no matter how carried away one gets with fabricated beauty, mankind cannot be content with imitations. While observing the decline of European civilization, I could not help wondering about the future of these remnants of traditional Japanese culture—flower arrangement and the tea ceremony—that seem to be flourishing so magnificently today.

Churches and God

I had never been to a large cathedral before. But when I visited the Padre Church standing alone in the fields on the plain outside of Milan, I was moved by a feeling that the spirit of Christ lives on there. I entered the dark church and, as I stood gazing at the figure of Christ on the altar in front, a priest came over. "God is not here, so you need not worship," he said.

"Now here's an interesting priest," I thought. "Where is he then?" I asked.

He told me that the church was an example of early Gothic architecture and had a long history. It was apparently rich in legend: A dove carrying a branch had alighted here, so forty monks settled and farmed the land. They had baked bricks and built the church by hand. Napoleon had even occupied it at one time. Because it had none of the stained glass windows and other embellishments which are so common in cathedrals, it seemed somehow to exemplify the spirit of Christ.

This priest showed an interest in natural farming and so we got to talking. We sat down on the lawn in the center garden and he split open a watermelon and treated us. Afterward, as we were leaving, he gave us some spices he had grown in the garden.

What delighted me most was to see a figure of Christ here like that which I had imagined.

At any rate, wherever one looks—the quiet, stone-paved roads, the sculptures on street corners, the beauty of the fountains, the stone castles in the forests, the crosses blazing with the rays of the setting sun, the beauty of the stained glass windows—European towns are filled with enduring reminders of its history. To the Japanese visitor, the Europeans enjoying a life of serenity in the midst of all this splendor seem very fortunate indeed.

But when one stands before the solemn statue in a square somewhere of a great king astride a horse, or before a sculpture by Michelangelo or some other great artist in front of a cathedral, one recoils before the force and power of these. This no doubt is because it is so plainly in evidence here that all European culture was created by man.

The constant struggle between God and man and between the rulers and the ruled is brought into bold relief in the churches and castles. I was appalled to find that the emblem of the castle in Milan shows a large snake (the king) swallowing the people. The name of one quiet country town west of Milan literally means "town of the dead." The people here were killed in an uprising and the memory of the tragedy preserved in the name.

In the courtyard of one famous cathedral, I found some words inscribed inside a circle two yards in diameter. When I asked what this was, I learned that a country priest who had called for a religious revolution had been burned here at the cross. I felt deep indignation at the idiocy of men of religion who, rather than lamenting over the wretched side of Christianity, strive to protect the dignity of God and enclose God within religion.

I felt a far greater sense of pleasure in the squares in front of Austrian and Italian churches where flocks of pigeons gather and booths sell picture postcards and local souvenirs, and at the open air markets in Paris where vegetables and fruit are sold. Perhaps my sense of relief at such places had something to do with the people free of ceremony and pretense milling about under the sun. Here, people can be renewed again.

For that reason too, it seems fair to say that the decline of European civilization arises from the Western philosophy of placing oneself first; that in trying to protect one's person with a stone castle, Europeans have ended by enclosing the self within a prison.

A Green Peace

The International Nature University ───────────

I skipped around Europe, getting a look at both its bright
and dark sides. At Paris, the last stop on my itinerary, I saw
what appeared to be a hopeful development: a "green
university."

A fellow there, the son of an important American Indian
chief, was practicing natural farming. He had asked me to
take a look at what he was doing, so I paid him a visit. He'd
taken my advice by scattering a mixture of vegetable seeds in
his garden at home, and was in effect practicing natural farming in the dead center of Paris. The first thing that he did was
to take me over to the Palais d'Elysée. I had no idea why we
were going there so I asked him what he was up to. "You
said there are no trees in Paris," he said. "So I'm taking you
to a place where there *are* trees."

When we got there, I found that not too far from the Palais
was an area thick with trees—almost like a virgin wood. In
the center of this stood a building that looked as if it was a
mansion built for nobility. I thought that he worked here
perhaps as a watchman, but it turned out that he was the
caretaker. "This," he told me, "is one of the proposed sites
for the International Nature University. I want you to decide
whether the University should be located here or not."

Now, I am no fortune-teller. I have no special insight into
such things. But when you are told something like this, you
have to give some kind of a response.

"In Japan, people often build shrines and temples where
the trees are large and the soil is black. They favor places
where the earth is good and rich. I see here four or five large

trees with diameters of more than two yards. The mere fact that these trees are here means that this place goes back at least a century or two. And although land in France is generally depleted, the soil here in these woods is dark and rich —what we call "bear-colored" in Japan. Plus you've got another twenty acres of what looks like virgin forest in the back. So I'd say that this seems to be an ideal spot for your school, yes."

He was delighted. We then went deeper into the woods, where he had erected an Indian tepee made of a heavy cloth stretched over a wooden frame. He said that he sometimes slept here.

"What's your sign?" he asked all of a sudden.

"Aquarius," I answered, wondering why he wanted to know.

"Touch this pole over here."

"What will happen when I do?"

"That will become your seat," he answered. "The university will be built by twelve holy men. This pole is a constellation. The place that you touch will be your seat." After I had done this, he continued: "Now you are a founder. Whenever you visit, you will stay here and give lectures or do whatever you please."

Before I knew it, just by touching a pole, I had become a founder of the university. This had all the charm of a fairy tale.

Afterward, he gave me a tour of the mansion. The rooms were of opulent decor, but most impressive of all was a spacious hall with mirrored walls and a large, polished, round table.

"What use does a hippy university have for such a hall of mirrors?" I asked.

"Someday we may invite the heads of state of five or six European nations over for a summit meeting here. We've got to have a room like this for such occasions."

"Today, it looks like a meeting of the kings of hippies," I

said in Japanese. The young lady interpreting for us broke
out in peals of laughter.

Some time after returning to Japan, I got a letter from him
and learned, with some surprise, that the school had gotten
off to a fine start.

When I inquired, I found out that this fellow was a Dr.
Norman William, who had some influence at UNESCO.
"Surprising things often happen when least expected," I said
before we parted. "I'll hope for the best without raising my
expectations." Again the interpreter laughed, but she refused
to translate what I had said.

Michio Kushi, who headed the summer camp at Innsbruck
and is a world leader of the macrobiotic movement, also
unfolded his ideas for an international university one evening
over supper. He had already started putting his plans into
action, having purchased a former Franciscan monastery for
the school. He even showed me photos of the place. Kushi
said that today's schools are no good, which is why he's trying to set up a totally new type of university. The name of the
school will have the word "nature" in it.

Won't you join in the effort?" he asked.

"A Japanese farmer doesn't have that kind of time to
spare," I replied. "But I'll be happy to give whatever advice
I can on farming."

With plans of this sort under consideration, it does seem
to me as if new higher institutes of learning critical of today's universities may soon emerge around the world.

The Paris Peace March ─────────────────────

An antinuclear peace march that had started out from Oslo
reached Paris the same day that I arrived there from Holland.
The march began in Oslo when four women got together and,
deciding that they had to do something in the name of peace,

simply started walking. As they proceeded south, people in the countries through which they passed joined them in their march. By the time they arrived in Paris, the number of marchers had swollen to an estimated 8,000 people.

That evening, the young Japanese woman who was interpreting for me said, "Several holy men are here from India. Would you care to speak with them?"

When we went over, several thousand marchers had already arrived and were sitting along the streets, speaking noisily with each other. We entered the auditorium, which looked like a large, renovated warehouse, and found that a jazz performance was underway. I suppose this was a way of thanking the marchers for their participation. The interpreter walked right up to the conductor at the podium and whispered something in his ear. At once the music stopped and she beckoned to me. The people in the hall began to clap their hands and yell, "Speech, speech!" I came totally unprepared, but taking heart from the good-natured kindness of these people, I stood on the podium and spoke for a while on the roots of peace and war.

Usually, I speak rapidly, rattling away at a furious rate. But because this was being translated, full of laughter-provoking blunders, by three interpreters into English, German, and French, I had time to think and was able, it seems to me, to get across a more serious message. When I finished speaking and stepped down from the podium, I was embraced by a delighted old man who was introduced to me as an Indian holy man.

I've completely forgotten what I spoke about, but that really doesn't matter. It was enough merely that people of different countries should meet, embrace, and share with each other the joy of being alive.

Weapons . . . for Defense? ─────────────────

In Europe, all the countries share common boundaries. There is no way of knowing when enemy tanks will cross the mountain passes at the border and enter one's country. That may be why none of these countries have become absurdly rich. I heard it said somewhere that "instead of defending one's country with weapons, it is better that each individual acquire a philosophy of firmly protecting peace." The notion of protecting the country and of protecting nature from destruction is very well established in Europe.

Perhaps this is why I didn't see any campaigns in Europe to plant trees. Nobody over there is saying to plant or not to plant. No one cuts the trees down, so things can be left alone. There are trees almost too big to get ones arms around spreading their branches out in the fields. In Japan, such a tree would soon set farmers quarreling. The fact that these are left alone to spread branches out to the edges of the fields tells me that the European way of thinking about and protecting nature differs fundamentally from that in Japan. To the Japanese, protecting nature does not mean protecting the trees but protecting oneself. With all the excitement lately over pollution or whatever, the word "protection" has begun to pop up everywhere. But the real meaning of this word has not yet sunk in. This word "protection" has arisen because nature has been decisively destroyed.

Having built high-speed cars and developed brakes, people are now yelling out slogans to drive safely at the top of their lungs. In the same way, instead of making brakes, it would make a lot more sense for the Japanese to walk rather than drive whenever possible, using bicycles for longer trips, as the Europeans do.

*   *   *

Some time ago, a group of five or six foreign visitors to my farm insisted that I go along with them to Hiroshima for the commemorative ceremony held there each year in August, so for the first time following the war I visited Hiroshima.

On our way back to Shikoku after the ceremony, one member of the group remarked that all the Japanese do is talk. "They speak without commitment," he said.

"What do you mean?" I asked.

"Why they're just giving lip service to the antinuclear cause. It's quite clear if you look at their attitude. While we were walking in Hiroshima, we noticed that young people don't ride the streetcars that go by. They'll take a taxicab instead, just to go a quarter-mile or less. There's no greater waste of energy than that."

I slept with the others in a small park behind the Peace Memorial Park. Children came here the next morning for radio calisthenics. Watching this, the other members of our group laughed at their lack of enthusiasm. "It's clear that these kids have come only because they want badges for participating. They're just pretending to go through the routine."

The ceremony commemorating the dropping of the atomic bomb is conducted in a festive atmosphere, but as soon as it is over, city officials come to clean up and put everything back in order again. Even though people are still talking, they try to get everyone out so that they can clean up. "Go on home now," they tell people who have come from far away to attend. Why have these people come? Why is a ceremony held in the first place? The master of ceremonies and the university professors spend an hour or more giving opening salutations and progress reports, but foreigners from fifty or sixty nations are only given two minutes to speak. I haven't the slightest idea why they bother to have such a ceremony in the first place.

At the ceremony, all they do is to proudly report that the auditorium is full or that this year's rites have been attended by so many thousands of people. The organizers put their entire efforts into things that don't matter at all, and don't seem to have any time to listen to sincere appeals from abroad. This explains why, in the boat on our way back, the foreigners with me kept saying that Japan is finished. Two were Dutch girls and one was a young man from somewhere in Europe. There was also an American girl and a farmer from New Zealand.

When I asked them what they thought of the ceremony, they answered, "We were totally disappointed. It wasn't at all what we had expected it to be."

Thoughtful Europeans believe that Christ's words, "He who lives by the sword shall die by the sword," are an absolute, unassailable truth. But the Japanese persist in the simple-minded belief that the more weapons one accumulates, the more securely one can protect himself. Weapons for defense strike fear in oneself and others. The mentality behind self-defense is out-and-out aggression. Defense and offense may appear unlike, the shield and the halberd may seem different things, but these are as joined in purpose and function as the right and left hands. One might as well say: "Stroke with the right hand and strike with the left." That's why, in Japanese, the characters for halberd (矛) and shield (盾) together mean contradiction (矛盾). But that is only the start of contradiction. The more we talk of offense and defense, the more these escalate, until everything gets out of hand. It would have been much better had we gone at it naked instead, like sumo wrestlers. The more powerful the arms we fight with, the more intense our wars. By building increasingly monstrous weapons, we are only creating the cause of what could be a monstrous tragedy. When people first began making weapons, they used these and were able to control them. But today, robots are building an endless supply of weapons and exerting control over people. Someday robots may use weapons and manipulate computers to provoke war. Wars arise from unexpected causes.

Today weapons are in a race by themselves. Things have gotten to the point where weapons are moving people. Adults may think that those invader and star fighter shows are just pure fantasy, but these are thoroughly inculcating our youngsters. Oh sure, it's okay if the film companies, toymakers, and publishers make money off unsuspecting kids, but while we're indulging in such self-deception, what are these kids who have been brought up so thoroughly on military pablum like this going to start doing? You can be sure that no one is going to accept responsibility for the consequences, whatever they may be. Today in Japan, all we're doing is clamoring over our individual liberties, about how people do as they please and nothing can be done about it.

Reflections on the Japanese ──────────────

After traveling around Europe, I got to thinking that the time has come when we must begin to seriously examine the nature of the Japanese. I saw the shortcomings of Westerners, but very often those same shortcomings, if you turn them upside down, can be great assets. Westerners have a strong ego. But because of this, they are capable of being completely dedicated to and consistent with a philosophy of self-love. They do not become wrapped up in a group and lose a sense of self, but retain always responsibility for themselves. This same self-love gives them a sense of independence that enables them to protect their own road and carve out a new path.

The failing of the Japanese is that, because they have been brought up in a home full of affection, they become obedient children who comply meekly with the desires of their parents. This has the drawback of fostering an irresponsible personality that is only too willing to leave things up to others. Thus, on the surface, the Japanese appear to be attentive to the needs of others, obedient to society, and cooperative. This mentality is reinforced by a host of sayings in Japanese, such as "The nail that sticks out shall be hammered in," and "One cannot win over one's betters." Such notions appear to be the key to how the average Japanese makes his way through this world. But should this pent-up anger explode within the group, he falls into step behind someone's banner and becomes capable of doing anything at all. "If everyone does it, it's not frightening," he'll say. This results in the carefree merry-making that occurs during the cherry-viewing parties in the spring, and in the wild, reckless flings that people will indulge in without the least compunction or concern for what others think.

The Japanese are a good people who love chrysanthemums. But, depending on the leadership, they can be transformed with great ease into barbaric warriors. Both they and others recognize them to be a courteous people of high-minded mentality and deeply religious sentiments, but to what extent is this true today?

Confusing religiousness with visits to temples and shrines, the Japanese today take part in magnificent religious events that have become tourist attractions. As the Japanese public submits to the illusion of a resurgence of religion, behind the scenes, men of religion are busily amassing money and power.

For their part, the philosophers too are noncommittally laying out plausible-sounding arguments, founded on a pluralistic set of values, that circumstances must be appropriately weighed. But wasn't it the mission of the philosopher to ask what pluralism is, and to unify the discordant value systems of our world? There can be only one absolute value system, and it should have been up to philosophy to show this.

What I fear most now is that, although they had religious surroundings and sentiments, aside from a very few individuals, the Japanese never had a firm religious core.

When the integral blend of Shinto and Buddhism which had developed in Japan was rejected as an illegitimate mix of two distinct religions, this was radically transformed and split apart without the slightest objection on anyone's part.

People nod in agreement and accept almost unthinkingly

absurdities, disseminated as reasonable statements, to the effect that Shinto is not a religion at all but just a set of folk beliefs and customs. It is freely used and abused in this manner by the state and the powerful. In a sense, Christianity, Islam, and any other religion would have suited the Japanese just as well as Buddhism, which goes to show that the Japanese themselves never had any religious convictions to begin with.

Although it is possible to take this as meaning that the Japanese people harbor the potential for aspiring to a religion that goes beyond religions and sects, the truth of the matter is that, quite to the contrary, the Japanese have become caught in a quagmire of heresy from which they can no longer extricate themselves. Perhaps it is to be expected that the ethics and morality of an irreligious people float about unmoored like waterweed on a pond, spinning and tumbling without end.

Japan today is rapidly being Americanized and plunging into a culture of decadence. What will come of it if it continues to follow in the footsteps of an aberrant America where news of a 50-percent divorce rate and 50,000 child abductions a year is normal. Because Christian spirituality is still very much alive in Europeans, they are more resistant to Americanization, yet the effects are apparent there too.

Japan is looked down upon by people all over the world as an economic animal, a materialistic nation of worker bees. But what are we to make of the Japanese, who fail to understand what all this means?

Look at today's scientists who have become the palanquin bearers of rampaging science. They dance about wildly, drunk on the sound of the words "high technology."

Look at the dangerous politicians who rush ahead madly along the road to the military-industrial merchants of death while hoisting high the flag of peace. And look at the public which supports this.

A Message for Peace

Almost fifty years ago, I wrote a letter to Franklin D. Roosevelt. I didn't get an answer, of course, but today I'm thinking of writing a similar letter again. The year was 1941. The Japanese military was invading China and Roosevelt was trying to check its advances. He even threatened to shut off the oil routes if Japan didn't take its hands off China. Rather than sitting around and waiting to die, the Japanese military insisted on fighting things out. The country had its back up against a wall.

Still a young man in my twenties back then, I felt that if I appealed to the American president's sense of prudence by writing my thoughts on the dreadfulness of the approaching footsteps of war and on the contradictions and errors of human judgment this entailed, that, however slight the chances, this might help avert the outbreak of war. So I wrote a letter to the President and placed it in the care of a reporter at the local newspaper in Kochi. Yes indeed, I was young back then. Of course, the letter was never delivered.

I view with sadness this day on which, after having reached my advanced age, I prepare to write down again the same thoughts that I expressed so long ago. Well as I am aware of its futility, I write this letter as if it were my last. I wonder whether the leaders of the United States, the Soviet Union, and other countries would care to consider the thoughts and feelings of an old farmer.

Like my first missive long ago to Roosevelt, this one too is written as an open, interrogative letter.

(1) Christ said, "He who lives by the sword shall die by the sword." I wonder if you think this was just the fervent desire of a dreamer?

People say that today you brandish nuclear weapons in your right hand while with your left you ship food strategically throughout the world. Do you believe that you can, with such a stance, become a champion of justice who saves the world? Do you think that you can gain the world's trust with weapons? In the East it is said that no grudge runs as deep as one over food. If you use food as a strategic weapon, you may earn the hatred of the entire world. You demand today that Japan liberalize its food markets, but how well do you sense the rage and indignation of Japan's farmers?

Isn't the very thought that peace is maintained in the world through force and strategems the empty dream of a fool? It is the sad destiny and doing of man to believe that the only course is to take an eye for an eye and to answer hate with hate.

Peace cannot be won with weapons. This has always been true—just look at history.

(2) I'd like to point out that a way of thinking based on the theory of biological evolution, which holds that in our world the strong feed on the weak and takes for granted that, according to the law of natural selection, the fittest shall survive, is founded on error and fallacy.

Nature in its original form is neither a world of competition nor of cooperation. Viewed macroscopically, in the natural world there is only the principle of a food chain about which revolves the animated flow of life and motion. I believe that the true state of nature is one where there are no strong or weak, no eternal victors or absolute losers.

That is why it is unpardonable to mistake natural selection as a law of nature and, believing the survival of the fittest to be the proper and universal course for human progress, to think it only natural that one must endeavor to build up military force and to win out over others and survive.

War exists only in man's world and is alien to the rest of the biological kingdom; it does not belong to the natural world. War is an absurdity that arose from the human intellect.

There are no grounds anywhere, anytime for justifying war.

(3) Man today, as in the past, wanders constantly between peace and war, forever exposed to the terror of hostilities. The trigger of war may be drawn on the basis of distinctions in false judgment that encompass the interests of the state, good and evil, love and hate. But where do you yourself place the standards for right and wrong in these human passions?

An immutable, unchanging standard for judging good and evil, right and wrong, that holds for the entire world is not something that can be arrived at through the human intellect. Why, man does not even have an absolute scale with which to measure wealth and poverty when it comes to resolving conflicts of interest.

I believe that the final judgment rests with God, that we have no alternative but to observe the laws of nature. How about it?

(4) How do you view the responsibility of leaders?

History attests to the fact that, while people throughout the world desire peace, wars are always entrusted to the discretion of a small group of people called the the leadership. Wars are never caused by the poor who occupy the bottom rung of a nation's social ladder or by the "stupid" peasantry.

Most people count on the shrewdness and farsightedness of their leaders, but do you not think that the discriminating knowledge of humans is nothing other than unenlightened knowledge and cannot become a means for resolving contradiction?

As human knowledge deepens, things do not become clearer; this only deepens the mysteries and increases the level of confusion.

Confusion grows and the clouds of suspicion spread as leaders carefully deliberate and spin out plans and strategems. Suspicion turns to fear, and fear breeds evil.

This is because, no matter how much reasoned caution is exercised and how many intelligent meetings are held, it is not possible to hold in check the passions that demand a tooth for a tooth and an eye for an eye. Nor is it possible to stem the endless competitive spiral of weapon for weapon.

Mankind has developed through the human intellect and now through that same faculty is in the process of collapsing. Yet, even at this juncture, we continue to seek salvation through the intellect. Can you not see that the only road to such salvation is by discarding the intellect?

The only option open to us now is to live by the wisdom of Buddha which transcends human intelligence and by the great love of Christ which transcends human love and hate. Never has this been more true than it is today.

We can no longer measure tomorrow with the intellect.

I feel strongly that the time has come when leaders must be saviors.

(5) In our materialistic world, it seems to me as if people rush about frantically solely in pursuit of economic affluence, as if they struggle with each other only over

conflicts of interest. This must surely result from the conviction that material plenty is linked directly with human joy and is the source of happiness.

Even though anyone would surely prefer to be an apostle for peace than a merchant of death, why is it that your country is throwing itself body and soul into the hysterics of an unstoppable nuclear arms race? Is it because, rather than becoming a powerless servant of peace, even if it means joining in conspiracy with the merchants of death, you favor the road of a powerful state and a supreme ruler. If so, this is a gravely perilous gamble for mankind.

True human joy does not come from material possessions; nor is happiness conferred from without. Such a path only leads one away from God.

To a heart filled with pride, the fruits of the mountains and seas served at a palatial feast are tasteless. Yet even Solomon, robed in his greatest splendor at the height of his glory, paled before a single lily in the fields. Ancient wisdom teaches us that, even without a single possession, we can be happy as long as there are flowers in the fields and birds singing.

It should be pointed out that the ideology of national affluence and powerful weapons is the greatest enemy of a nation's people because it robs them of true joy and happiness. When a country grows affluent, the people become proud and decadent. When the military grows strong, it gains influence and robs the people of their freedom.

The truly wise and courageous do not need wealth or weapons. On the other hand, no matter how much they amass wealth and how high they build their castles, those whose hearts are not generous cannot enjoy an easy sleep.

The more a country protects itself, the weaker it

becomes. The more it presses forth in pursuit of the enemy, the stronger becomes the enemy and the more remote become peace and freedom.

(6) I wonder what you think is the key to politics. The sages of the Orient say that this is the road of moderation, which could be thought of as standing at an unmoving center.

From the standpoint of Western religion and philosophy, ours is a world of contradictions and the key to resolving these is to walk a middle road while balancing right and left. This is why meetings and discussions are so common in the West and so much effort is devoted to seeking harmony and accord. However, the middle road traced by a dialectic process such as this is not the road of moderation to which Easterners refer, but just a half-way street that leads nowhere.

According to Oriental philosophy, ours is not a world of relativity but an absolute, selfless world that transcends space and time. Oriental philosophy holds that the standpoint of absolute universality—which may be called the standpoint of God—known only to those who stand at the true void transcending space-time is capable of becoming the Great Way of politics.

Any attempt for dialogue and cooperation from a relativistic perspective will only make the road almost impossibly long. But if one stands himself where God stands and opens his heart, consensus and dialogue will no longer be necessary. In Rome and London, people would not themselves commit the folly of closing down the natural sea lanes. Instead of threatening them with weapons, why not invite all the people of the northern and southern hemispheres over to America sometime? This will prove that Tolstoy's Ivan the Fool, Gandhi, and the farmers of America are also part of one unity.

Surely then, people will see that what they thought of as the enemy was their own shadow, and that contradiction is nothing more than two sides of the same reality.

The swords of demons cannot stand up to the hands of babes. Weapons raised against naked children are all the more dangerous. The greatest weapon of all may well be the bare hand.

In Europe and America today, surely the time has come, not to decide between guns and butter, but to follow the spirit of Christ and think seriously of what it is that people need to live.

(7) I am troubled today by a great anxiety. People are hurtling forward into an age ruled by materialism and exaggerated faith in science, and in the process are metamorphosing into sophisticated computers.

Believing the basis of life to lie in the genes, life scientists have analyzed these and discovered that man is nothing other than a vehicle for transmitting genetic information.

An age in which parents and teachers worry over the ideologically prejudiced education of children is still preferable, for what is being attempted next is the handling of all problems and concerns by computers and robots in place of humans.

Is there someone today with the courage to be the first to press the nuclear button? Since no human could possibly have any longer the ability to assess all the relevant information and make an absolutely correct decision, the responsibility will have to be delegated to computers and robots. But will machines arrive at decisions superior to those of men? Computers are not capable of becoming anything more than faithful watchdogs programmed by their masters. Yet the day will come when stacks of data processed and output by computers will be transmitted directly into the human brain, which will have become a mere data receiving unit, in order that computers are able to give orders to humans and manipulate them.

Also of concern is the rapid progress being made in biotechnology. Scientists who haven't the slightest idea of the true meaning and purpose of human life are busy devising ways to program life.

I have reason to believe that new forms of life created with computerized data and genetic engineering will become a source of human calamity. Here is why: Kant says that human thought and man's various basic notions are all erected on the universal *a priori* forms (notions) of space and time. Naturally, computers too are only devices constructed on the basis of the concepts of space and time. Hence, they are merely machines which make erroneous decisions that differ fundamentally from the decisions and aims of God, who makes decisions from a standpoint that transcends space and time.

Moreover, no matter how superb the living things created by life scientists through genetic recombination, these are created with efficiency foremost in mind. Scientists cannot possibly create perfect organisms that go beyond the vast and infinite standpoint of God.

Scientists today are under the illusion that they can replace God as the Lord of all creation. But no matter how much they may squirm and struggle, there is no getting around the fact that the supermen and new organisms created by man shall forever be imperfect creatures caught within the domain of the concepts of space and time.

My concern is this: In the event that, foundering under a deluge of information, you were neither able to escape the grip of the merchants of death nor check the wild rampage of science, who should people look to for succor? Having deserted God, man will not be able to return to His side. Does this mean, I wonder, that humanity will become an orphan of the universe?

Are my concerns nothing more than empty fears? Is my thinking perhaps all wrong? If you could tell me, it would be a most unexpected delight.

Food and the Ecosystem

The Japanese Diet Takes a Turn for the Worse

I think most people would agree that, judging from the structural features of the human body, the proper food staple for man is grain. Although we all have cuspids for biting and tearing meat, the rest of the teeth we use for chewing are molars. The human skeleton and the structure of the human jaw also make man suited to the eating of grains.

Japanese peasants in particular have retained the diet of a herbivore. Given the country's climate and terrain, this diet has consisted of grains and vegetables. In terms of calories, rice, barley, foxtail millet, proso millet, and barnyard millet —known in Japan as the "five grains"—long served as the staples. Consumption of the five grains was a basic dietary principle in the traditional Japanese farming community.

Today, however, little of these five grains is grown anymore, except for rice and barley. The minor grains have been all but forgotten. With their small seeds and low yields, production of the millets has shifted over almost exclusively to rice and barley. But it seems to me that it is the primitive, small-seed crops closest to nature, such as the millets, the *azuki* bean, and buckwheat, which are best for human health. So I would say that, of the five grains, those with the small seeds that are the most primitive are probably the best to eat.

In our modern world, however, everyone yearns after what is large and tasty. "Bigger is better" seems to apply today to fruits as well as grains. Yet, since energy is more highly condensed in smaller things, I would venture to say that it is probably better to go with what is smaller than bigger.

The diet of the Japanese farmer has undergone rapid

changes. Today the minor cereals are disappearing and the word "vegetables" is being replaced by such terms as "leafy vegetables" and "greens." Meanwhile, the quality of what is being grown is deteriorating; rice and barley today is no longer the rice and barley of yesterday.

The Japanese are in the process of forgetting more than just the five grains and the "five vegetables." As I have already pointed out, there was at one time an enormous variety of foods in Japan. By adding together the grains, vegetables, fruits, and edible wild herbs that can be gathered from the fields and hills about my village, one can create a food mandala of the four seasons.

It is said that the farmers of old were poor, eating only rice cooked with barley, pickled vegetables, and the sour pickled plum (*umeboshi*), but in a sense their's was a marvelous diet.

Try as one might to gather together the best out-of-season foods from the supermarket shelves, one cannot create from this a true feast. Nothing better characterized the land and diet of Japan than its abundance of vegetables and edible herbs.

Too Few Vegetables in the West ———————————

While in Europe and America, I was surprised at the small variety of vegetables grown and eaten there. About the only root vegetables that Europeans and Americans seem to eat in any quantity are carrots. There is plenty of burdock growing wild, but no one eats it. They don't even think of this as a food. I remember hearing about how Japanese soldiers who fed Americans burdock at prisoner-of-war camps during the war were accused later of war crimes for forcing prisoners to eat "tree roots." That is the extent of American ignorance when it comes to root vegetables. The same is true in Europe as well. They do grow potatoes and some other tubers, but these are very hard and not what I would consider edible fare at all. I did occasionally see sweet potatoes, but they were quite different from the delicious, high-yielding sweet potatoes common to Japan. And no one eats taros, no one eats burdock, no one eats lotus root or bamboo shoots.

The fact that Westerners eat leafy vegetables but almost no root vegetables means that they are eating only half of what they should. Even in terms of nutritional balance, the deeper roots penetrate into the soil, the more primitive they are and the better for the body. The yams ought to be the best. By eating root vegetables good for the body and vegetables high in fiber such as burdock, sweet potatoes, and bamboo shoots, one will never be constipated and will enjoy the beneficial effects these have on the health. But instead of these foods, people in the West eat only meat. I'll bet that this is why so many people are constipated. If I may be pardoned for waxing poetic again:

> Sweet potato turds,
> Left behind, make fit and strong
> The touring pilgrim.

When I was in Europe, I talked to people in Italy and everywhere else I went about sweet potatoes. Although they could grow sweet potatoes quite easily there, almost no one seemed to know about this vegetable. Once I even began a talk to a group of local farmer's wives by telling them about my special high-yield method of sweet potato cultivation. In any case, there does not seem to be any other country with such a rich variety of foods for cooking as Japan.

Returning a moment to rice, barley, and wheat, these are the best crops in terms of the calories of food that can be produced on a given acreage. The calorie outputs of corn and potatoes are also high, however, in colder climates. In Japan, growing rice and barley or wheat gives the highest output of calories, and cultivation is easy. Hence, these crops are the best choice when it comes to making effective use of limited land.

To this day, I have fought against the tendency in Japan

following the war to drain the rice paddies and plant the land with citrus groves. The Japanese farmer must not stop producing rice, barley, and wheat. Not only do these grains give the best yield of food energy from the limited arable land, they are the most appropriate for Japan's climate. Another reason is that man draws the great bulk of his calories from rice and wheat. Roughly half of the world's population uses wheat as its food staple, while the other half uses rice.

The nomadic peoples of the cold northern countries hunted and enjoyed animal flesh. They became meat-eaters to make up for the shortage of grains. In Europe and America, the practice arose of using wheat, which grows well on cold, dry, elevated ground, as the dietary mainstay, and eating also a lot of meat. Once it was ground into flour, processed, and made into bread, wheat was a convenient food. It spread first among nomadic peoples because it could be preserved and was convenient to carry. In the warm, humid subtropics however, upland rice was grown in hilly, elevated locations, and paddy rice on flat land.

Although mankind can be divided into rice-eating peoples and wheat- and bread-eating peoples, the questions of whether man should make rice or wheat his staple and what he should eat in general, including assorted cereals and vegetables, must be settled. All man had to do was to follow the course of nature and feed randomly on the foods that arise naturally in the region that he lives. The question to ask is not what to grow here, but what grows here. Will it not suffice to take as the staple food for winter the rice harvested in the fall, and to take as the food for summer the barley gathered in late spring?

The Japanese farmer did not ask what food to grow and eat. He simply harvested and ate what grew naturally in his fields in each passing season. And he cooked this food in accordance with the principles of nature. There are not first people who grow food through the devices of the intellect;

first there is natural food and people who live in a "do-nothing" nature.

This was the original way of living in Japan and the proper way of eating (a sacred act). But, unfortunate as it may be, things have changed completely over the past decade or so.

Japanese Diet and Cooking Are Disappearing ──────

If I may say so, no other people originally had such a sharp sense of taste and were as good at cooking as the Japanese. One strong feeling I had throughout my travels in Europe and America was that Japan's cooks and chefs have no need to go elsewhere for instruction. All they ought to do is visit a Japanese farmer's wife. I felt certain that the skill of the farmer's wife in flavoring food would allow her to pass for a master chef anywhere in the world.

I would even venture to suggest to those people studying Western cuisine in Japan that, rather than going to all that trouble in Japan, they would do better to go abroad and provide culinary guidance there. One reason is that, although solid progress has been made in Japanese cooking, this has caused it to diverge from nature, resulting in a decline in the quality of the materials. Even though the skill of cooks has continued steadily to improve and the number of Japanese chefs worthy of international acclaim has increased, they no longer have good materials to work with. The only materials available today are processed foods prepared chemically from petroleum products. That includes everything from vegetables such as tomatoes, eggplants, and cucumbers, to rice and wheat, and even fish, which are the product of fish farming operations rather than caught fresh from local waters as in the past. Either that, or trawlers bring back deep-sea fish from the South Seas. No matter how skilled the cook, if the materials he works with are no good, there is nothing he can do.

That being the situation as it stands today, I get the feeling

that things may reverse. While the Japanese have Oriental
bodies, their thinking is being polluted by Western philosophy; they are becoming devotees of science. Scientific agriculture is being practiced today in Japan, and the foods
eaten here have become Western-style petroleum-based
products. Given their dogged preoccupation with nutrition at
the exclusion of all else, the Japanese too will surely become
a meat-eating people.

Young people in Japan today have taken to eating ham-
burgers at standup counters. But is it really all right for them
to be eating that kind of instant food? Clearly, a confused
diet has misled people and, once misled, they begin to adopt
sumptuous and extravagant eating habits. From this high
point, they will probably plunge downward at an accelerating
rate. Westerners are already aware of this. The foreigners who
come to visit my farm in Shikoku tell me: "Frankly, I'm disappointed by what I've found in Japan. I thought that because natural farming had been developed here, it would have
caught on well by now. But nobody's doing it over here. And
when I went to visit shops selling natural foods, none of them
begin to compare with the outfits being operated in America.
The average person in the street knows and cares even less.
People in general don't have the slightest inkling of the true
value of a natural diet. It seems that the only ones eating a
natural diet here are the sick and the nature-lovers."

A clear reversal seems to have occurred in the Japanese
attitude concerning essentials—food, clothing, and shelter—
about ten to fifteen years ago. I felt this keenly while standing
recently in front of the Yokohama Customs Bureau where I
worked close to fifty years ago. Working as we did in the
plant inspection division, we were government officials so we
could afford to look important, but we never felt ourselves to
be the equals of the foreigners there. There was a servility, a
baseness, about Oriental people. Westerners in their natty
clothes strolling through Yamashita Park in Yokohama with
their children looked so dignified and composed; as a race,

they seemed a cut above us. The children would walk right
into the fancy restaurants and hotels totally unperturbed.
Now that I think of it, that was perfectly natural. Japanese
children today enter these establishments without the least
hesitation. Way back then, Western children could strut
right into such places, but Japanese had a hard time just
getting in. Even customs officials such as us had a hard time
entering places like these without feeling self-conscious.

Lately, however, when I pay the port city a visit, I find that
it is the foreigners who are dressed poorly and the Japanese
who are all spruced up. In restaurants, even the children sit
calmly in their chairs and look up at the waiter as they place
their orders. When I see this attitude, I cannot help feeling
that things are now the exact reverse of what they were when
I worked here. The Westerners slip in quietly, eat something
plain off in a corner of the restaurant, and hurry out again.
The Japanese, on the other hand, come in grand style, and
even the children appear to look down upon Westerners.
That is how different things are today.

But somehow I get the feeling that we could see another
reversal again ten years from now. People in Japan used to
look in envy at the proud Westerners and the thick steaks
they were eating in the restaurants, and think, "Gee, I'd like
to try eating steak like that just once." Well today they are
eating steak, while Westerners have taken to vegetarianism.
This reversal in diet seems to suggest something.

Confusion over diet confuses the body and the mind. It
affects everything. The health of the body comes from the
diet. And thought arises from the body.

Diet and Thought

If one stops to consider what thought emerges from, it is
clear that if the diet differs, the physique differs and thinking
differs. Even the blood changes. A diet with lots of meat

results in acidic blood while a vegetarian diet makes the blood alkaline. Alkaline blood gives one a gentle, peaceful disposition. Oriental people are said to be mild-tempered yet warlike, but in reality the people of the Orient are quiet. Grain-eating peoples become quiet and peace-loving. But when hunting and fishing races eat meat, such as do the Westerners, their blood becomes acidic. The foods they consume are highly concentrated energy. To borrow a view propounded by George Ohsawa, founder of the modern macrobiotic movement, meat is yang. People who eat vegetables and fruit become yin. When one is yin, one becomes quiet and feminine. When one eats flesh and blood, the blood is cloudy and acidic. One is masculine, active, and aggressive. Alkaline blood makes one quieter, calmer, and more peaceful, while acidic blood makes one more assertive and masculine. This assertiveness helped the West conquer the entire world at one time. Largely vegetarian races—such as the Oriental and Negroid races—were suppressed and defeated. Meat-eating races are like lions. They show intense energy and are mentally advanced. Because they are knowledgeable and physically strong, they organized the Crusades and conquered the world. But they are unable to sustain their efforts for very long. When it comes to a marathon, the vegetarians and grain-eaters are superior runners. That is why, even though these latter may have appeared to be facing ruin, they were able to hold out and are today again reestablishing themselves. On the other hand, we have a people that tried to subjugate the world with physical strength but was unable to do so completely and has now stopped trying. I think this is what lies behind the aging phenomenon we are seeing in the West.

With the West at a standstill due to the rapid aging of society there, the Japanese have caught up to and passed it by. That is how far Japan's economic growth has taken her. Even as they are being passed, Westerners appear to stand by quietly with their hands folded. But with the weight of centuries of philosophical insight, they are looking on calmly and coolly. The people of the Orient, and especially the Japanese, have come far, gaining in importance and prestige to the point where they lead the world in many ways. But to Westerners, the Japanese are merely retracing the route passed over earlier by the West.

A Seed War Is On

America's Food Strategy

It seems to me that the glory of America is largely a product of the scientific methods of farming that have been developed over the past two hundred years. But today, the problems of pollution and entropy engendered by modern agriculture are starting to shake the very foundations of the American system. I believe that one of the things that is beginning to arise in the dark shadows of civilization is America's food strategy.

America today prides itself on its role as the bread basket of the world. President Reagen boasts that America leads the world in grain production, apparently believing that the country can lead and even conquer the world with food and weapons. For some time now America's basic plan has been to conduct these two major strategies independent of each other. Whether they are aware of this or not, the Japanese have complied handsomely.

But I have my doubts as to how long this strategy can be pursued. I believe that its low regard for the land will be America's downfall. If the land falls victim to American farming practices and continues to be ruined at the present rate, it won't hold out even fifty years more. It may last another twenty or thirty at the most. Indeed, continuing scientific farming for even twenty more years will be difficult. I suspect that most farmers will give up before then. The eve of this breakdown in agriculture will be frightening as circumstances everywhere become desperate.

What I fear most right now is that the land which has nurtured and protected America's farmers is in the process of being ruined. Each year, American agriculture is approaching a limit of no return. Once this is exceeded, no amount of effort will succeed in reestablishing the farming methods of the past. When the land perishes, willing or not, the defeated farmers will have no choice but to become the pawns of agribusiness. Food production will have to be carried out as part of a broader strategy.

Let me be more explicit. Control over America's seeds today lies squarely in the hands of five oil companies. Working hand in hand with political and economic interests, the oil companies have moved into bioindustry and begun to take control of agriculture. The seed war started long before this. The moment that America's oil companies consolidate their control over the seeds of cereal grains such as rice and corn and over superior lines of livestock, America's farmers will be done for. The truth is that the hands of the oil companies have already stretched out over the entire world.

Even I myself have been drawn into this vortex.

Rice Seed as a Weapon

For many, many years now I have devoted myself to rice production, believing firmly that farmers in any age must never forget rice. However, with the rice we have today, one cannot grow natural rice. Today's rice has been bred and improved so much by man that I find it weak and totally unsatisfactory. Long ago, I felt that if strong, hardy rice suited to natural farming could be developed, it would be easy to grow, so I set out half-seriously to create new varieties of this grain. My goal was the exact opposite of the new varieties that agricultural scientists strive for.

My ambition was to bring back the healthy rices of yesterday. I even thought that if such varieties were to be brought to countries suffering from food shortages and there grown successfully by natural farming methods, this might help to halt the encroachment of scientific agriculture. However, if such seeds were to fall first into the hands of the CIA, they would be crossed by the oil companies with the male sterile rice developed by Professor Choyū Shinjō (see article below) and soon become a hybrid rice used by the powers that be as a tool for profit-making and strategy. It is indeed sad that when countries start competing to develop hybrid rice as a strategic weapon, the whole effort degenerates into the very same sort of ugly confrontation as the nuclear arms race.

The following article, which appeared in Japanese in the January 3, 1984 issue of the national newspaper *Asahi Shimbun*, relates my experiences with the new variety of rice I developed on my farm and offers a glimpse of the international "seed war."

* * *

Japan, U.S. Vie in Development of High-Yielding Rice Seed

Masanobu Fukuoka, author of *The One-Straw Revolution* and advocate of a way of farming that does not rely on pesticides and chemical fertilizers, is worried. At his home in Iyo-shi, Ehime Prefecture recently, the seventy-year-old natural farmer held his head in his hands and, sighing repeatedly, kept saying how he had gone and done something that could not be undone.

What Fukuoka has done is to cross a glutinous rice someone brought back from Burma after World War Two with Japanese nonglutinous rice. After years of crossbreeding and selection, he has succeeded in developing many lines of superhigh-yielding rice. Based on calculations from the number of grains grown on a single square-meter of land, these new varieties yield about one metric ton of rice per quarter-acre, which is roughly twice the average yield in Japan today. He genuinely fears that these superhigh-yielding varieties could kick up a confrontation over seeds between American multinational firms and Japan.

In March 1983, Fukuoka applied to the Ministry of Agriculture, Forestry and Fisheries for registration of four of his new types of rice as new cultivars under the Seedlings Law. Late last year (1983), while on other business in Tokyo, he decided to check on how his application was going and paid the Ministry a visit carrying with him seedlings of the four lines.

"Don't let out a single seed": At the Ministry, four officials surrounded Fukuoka and told him, "If the U.S. gets hold of this and converts it into hybrid F1 rice, they'll turn it against us. We want you to refrain from sending this seed out of the country for three years and watch carefully any visitors." Each of the officials added words of caution, one going through the routine of pretending to pick up a seed and place it in his pocket, warning Fukuoka that "someone may even try to sneak off with just a single seed." What made an especially strong impression on Fukuoka was the remark: "Make sure this isn't a repeat of the Shinjō incident."

Fukuoka worries that his super-yielding rices may get caught up in the national drive by Japan to develop its own hybrid F1 seeds to counter U.S. hybrids. This confrontation between Japan and America dates back to 1981, when William Davis, counsel for agricultural affairs at the American Embassy in Japan, called upon Toshiaki Ashizawa, head of the Agriculture Production Division at the Agricultural Ministry. It seems that U.S. seed companies wanted to sell rice seed to Japan. The purpose of Davis' visit was to ask for information on the technical and legal feasibility of entry into the Japanese seed market.

Citing the highly particular Japanese taste for rice, the complex natural conditions in Japan, and the existence of both a Foodstuff Control Law which in principle requires permission for the import and export of all food-related items, including seeds, and of a Plant Communicable Disease Control Law, Ashizawa explained that entry into the Japanese market would be difficult. He also indicated that, from the standpoint of national security, the only sensible course for a country was to produce its own rice seed domestically.

In spite of this, Richard Samuelson, president of Ring Around Products, a major U.S. seed maker that commercially sells hybrid F1 rice seed, came to Japan in March 1982 and visited the National Federation of Agricultural Cooperative Associations (Zenno) and the Agricultural Ministry. He also approached several Japanese companies about forming partnerships for the production and sale in Japan of hybrid F1 rice seed. The Ministry told Samuelson the same things it had told Davis but, undaunted by this, he returned again to Japan that same June and stepped up his efforts to develop the local market.

This initiative to sell rice seed, and especially hybrid F1 seed, caught the Agricultural Ministry totally by surprise. The Ministry had not conducted any research or tests to speak of on F1 rice. Plans did exist to start the production of other rices for use in processed foods not requiring taste considerations. Although it had launched a 15-year plan in 1981 that called for the development of superhigh-yielding strains of rice, the Ministry had not included F1 rice in this. Funds were quickly earmarked for F1 development from the Ministry's fiscal 1982 budget, but had it kept a closer eye on the hustling movements abroad concerning hybrid F1 rice, the Ministry would never have been caught off its guard.

To understand why this happened, it is necessary to take a look at the episode involving Professor Shinjō of the Faculty of Agriculture at Ryūkyū University, whose name the ministry official brought up before Fukuoka. Shinjō was the first to demonstrate to the world the potential for the agricultural production of F1 rice (see below for an explanation of F1 rice).

This reporter visited the professor late last year at his

laboratory in the city of Naha on Okinawa. Over dinner, he talked of how interest abroad in F1 rice had risen sharply and how a number of companies had contacted him directly about the rice.

Many years ago, Shinjō came up with a breakthrough in rice breeding by creating a set of three complementary lines: a male sterile line with the required traits, a maintenance line that can be reproduced without disappearance of the male sterile trait, and a restorer line that, when crossed with the male sterile line, can be made to bear F1 seed by self-pollination. He made a presentation on this at a meeting of the Japanese Society of Breeding in 1966 and reported his findings in a genetics journal in 1969.

Shinjō was born on Ishigaki Island in the Ryūkyūs. Near the end of the war, his father was inducted into the local defense unit, forcing Shinjō, along with his mother, who died soon after of illness, and four siblings to scour the countryside for food in order to keep from starving. It was this experience that set him on the road to breeding rice. He studied at both Ryūkyū and Kyūshū Universities. While a graduate student, he found the genetic mechanism of rice that would allow male sterile lines of any desired trait to be produced, thus providing him with a clue to the practical development of F1 rice. His first experiments at Ryūkyū University were done using buckets in place of rice fields. Later, he even drew on his own salary for funds to set up test fields.

The first to notice Shinjō's results was not Japan, but China. Increased food production was an indispensable part of that country's efforts to rebuild itself. However, in Japan, which was vigorously promoting a rice acreage reduction program, Shinjō's research evoked no real response.

In the summer of 1972, just prior to resumption in diplomatic relations between the two countries, China sent an agricultural mission to Japan. While there, mem-

bers of the mission asked for complete sets of the male
sterile line along with the accompanying maintenance
and recovery lines developed by Shinjō. Only too glad to
oblige where this could be of help in raising food production, Shinjō spent two evenings lecturing to the Chinese
delegation at a hotel in Tokyo and handed over a total of
180 seeds consisting of complete sets of six F1 varieties.
The Chinese had been working on tests for the development of practical F1 rice using a male sterile strain
discovered on Hainan Island. The 180 seeds received from
Shinjō were immediately added to the project. In 1974,
Shinjō gave a week-long series of lectures, six hours a day,
in Peking, telling everything he could about his results
thus far. Then, in 1978, announcing that they had succeeded in their efforts to develop practical F1 rice, the
Chinese invited Shinjō over to see for himself.

From China to America: In 1979, the U.S. oil giant
Occidental Petroleum acquired the rights to the use of
China's F1 rice for ten years. Ring Around, the seed
company whose president came to scout out the Japanese
seed market in 1982, is affiliated with Occidental. This
company attempted to introduce the F1 line into Japan
exactly as it had gotten it from the Chinese, without any
further breeding.

Occidental later divested itself of the company without
so much as a word of warning, leading some in Japan to
question the true intentions of the Americans. It was even
speculated that the U.S. companies had simply been
trying to resell the rights obtained from China to Japanese
firms in order to help offset a business slump within the
Occidental group. But Tatsumi Ono, director of the
Association for the Protection of New Varieties, who
came into contact with the president of Ring Around,
said flatly, "They're not giving up. Far from it. As a matter
of fact, they're getting ready to start up operations over

here." Occidental is not the only major U.S. industrial conglomerate that has set its sights on commercializing F1 rice in Japan.

Although it is not likely to happen in the near future, should the U.S. eventually succeed in developing strains of F1 rice that are relatively well-suited to natural conditions in Japan, and should sales in Japan prove to be profitable and problems concerning plant quarantine surmounted, what then could happen? Efforts to keep out these F1 seeds with the Foodstuffs Control Law, for example, could aggravate trade friction between the two countries. But let us suppose for a moment that F1 seed produced in the U.S. were imported into Japan. Since the agricultural production of the F2 (second generation) rice is impossible (see technical description of hybrid F1 rice below), Japan would have to import the F1 seed every single year. As long as business is good, the company producing the seed will never part with so much as a single set of the F1 parent seed that is the lifeblood of its business. This situation will effectively place part of Japan's rice production capacity in the control of the seed companies supplying the rice seed, and the countries to which these belong.

The head of Pioneer, America's largest seed company, has reportedly said that F1 seeds for corn are being exported from America to the Soviet Union to help bolster grain production there. Naturally, in such a case, the American side is holding on tightly to the parents of that F1 seed and wouldn't dream of letting it go.

Even in the case of hybrid F1 seeds, the larger and more diverse the genetic pool available for breeding, the greater the influence this can help bring to bear on seed markets. That's because varieties that are genetically rather remote from each other tend to result in a clearer expression of traits.

With the exception of a very small number of experts,

Japan as a whole has shown little interest until recently in the collection of such genetic resources. Word has it that the United States and the Soviet Union currently have amassed the world's largest collections of crop germ plasm. The U.S. has a tradition of instructing those in its diplomatic service to collect plant germ plasm wherever they were stationed. The boat that Commodore Perry came over on, and even the U.S. Occupation Forces following World War Two, collected germ plasm in Japan.

Promising "Fukuoka F1" Rice: Professor Shinjō is now looking with interest at Fukuoka's superhigh-yielding rice because it contains the genes of an ancient Burmese rice that is genetically remote from existing Japanese varieties. Rice is believed to have originated in the region that includes the Yunnan Province in southwest China, Burma, and the Assam district of northeast India, which is why Shinjō firmly believes that outstanding gene plasm which has not been distorted through artificial breeding remains present in the native varieties there.

Such being the case, the professor cannot help dreaming of the opportunity to bring Fukuoka's superhigh-yielding rice to its fullest potential as a hybrid F1. Fukuoka has received letters from people in China and South Korea requesting samples of his rice seed, although the letters give no indication of the reasons for the requests.

Of course, there is no telling how things will turn out if experiments for F1 production are conducted on Fukuoka's rice. But should success be had in developing powerful F1 lines with these, it will mean that a couple of country folk out in Ehime Prefecture and the Ryūkyūs will have rescued the Japanese from their narrow-mindedness. Up until now, heedless of the fact that many countries have inadequate food production, the Japanese have frowned upon research on rice, saying that as a country it already produces too much.

Wary of Exploitation by Corporations: Fukuoka does
feel as if he would like to help a lagging Japan win out
against the designs of multinationals gunning for control
of the country's food resources. But he is troubled by
grave doubts over the present race to develop new
varieties. His fear is that, as a result of endless genetic
tampering to get certain desired traits, rice is being
degraded into a greenhouse crop that cannot survive
without massive applications of pesticides and chemical
fertilizers.

Fukuoka maintains that his rice is complete in itself
without being turned into an F1 crop. "Not only is it
high-yielding," he says. "It requires no pesticides or
chemical fertilizers." That is far more than can be said
for current varieties of rice.

Fukuoka is quite willing to hand over his rice if it is to
be grown under natural conditions in developing nations,
but he is uneasy about the possibility of others making
improper use of the seed. A recent Agricultural Ministry
memorandum made Professor Shinjō promise not to
release the germ plasma for his F1 rice within Japan for
five years, and outside of Japan for ten years.

Hybrid F1 Rice—What Is It?: The method normally
used to create a new cultivar is to cross two different
existing varieties. The seeds produced from this cross are
planted, and from this grows the first generation (F1)
crop. Although this crop shows striking characteristics
such as uniformly high yields, the phenotypes and strains
of the crop diverge in the second (F2) and subsequent
generations. By selecting for the desired traits over the
course of many generations, the parent and offspring of
a line eventually come to show identical traits. What
happens is that this becomes established as a potential
cultivar. The rice grown by farmers today consists of these
established varieties. That is why part of the harvested
grain may be used to plant the following year's crop.

If a farmer chooses to raise seed at home as in the past, then he cannot use F1 seed for agricultural production. But it may be possible for seed companies to gain control over the parent varieties and grow F1 seed in large quantity for supply to farms. Following the war, American corporations brought this method into practice for corn, which is used as a major feed crop. Today it has become a powerful lever for U.S. supremacy on the world corn market.

Although both rice and corn are self-pollinating, in corn, the male flower (tassel) is attached at the top of the stalk, separate from the female flower, so this may be lopped off (detasseled) by machinery and the plant crossed with other varieties. In rice, however, both the stamen and pistil are present in the small flowers, making it impossible to quickly remove the stamens one by one from the flowers for each grain of rice in an actual seed-production paddy field. In order to create F1 seed rice, it is thus necessary to find a variety of rice in which the stamen is sterile—that is, male sterile.

*   *   *

I named my new strains of rice "Happy Hill." The Agricultural Ministry told me that they would rather not use the word "Fukuoka" because this might be confused with Fukuoka Prefecture. Hence, because the characters for Fukuoka (福岡) literally mean "happy hill," I immediately chose this as the name.

Tatsumi Ono, who is familiar with the American hybrid rices in the news lately, examined my rice and said that no short-culmed japonica rice produces more grains per ear than Happy Hill. As for taste, I'd say that it is average—certainly adequate for practical purposes.

Beyond Science

Insects Also Create New Varieties of Rice ────────

I would like to turn now to a topic that sheds some light on the question of what man has really accomplished. Perhaps I can help offset the dismal ugliness of the world of human knowledge by providing a behind-the-scenes look at the process of natural development—namely, the fascinating part that insects played in the creation of my new varieties of rice. That's right, several of the new cultivars I have developed have been the products of a collaboration between me and the insects in my fields.

I became aware of this one day when I sat down on the levee in my rice field and was crossing rice using scissors and pincers. First I would cut off the top third of each grain on an emerging head of rice with my scissors and remove the six stamens with the pincers. The next day I would sprinkle pollen from another variety over the head and cover the head with a bag. As I was doing this one day, I noticed that the insects next to me were doing exactly the same thing. Grasshoppers, locusts, and crickets were biting holes in the soft hulls. These holes were of various shapes, some round, some not, but the insects were making the incisions more skillfully than I.

Early the next morning when I went to take a look at the plants, I found snails, pill bugs, slugs, and cutworms all over the rice heads. I realized upon closer inspection that in some cases, these had eaten the anthers. When a grasshopper opens a hole in the rice husk and a snail feeds on the stamens, the pistils are later fertilized by pollen from neighboring heads of rice. I don't know whether the pollen is carried over by the wind or by flies and bees. In any case, I learned that seed fruition sometimes occurs within seed coats that have been bitten open by insects.

Of course, in a normal field, seeds with torn coats stand no chance of survival until the following year. But my field has not been plowed, or pesticides administered here, for thirty-five years. Moreover, during the winters, this becomes a barley field covered with uncut rice straw.

So there is always a chance that grain created here by insects may ripen in the autumn and fall naturally to the ground, there to pass the winter hidden and safe. This single grain of rice may germinate and grow into a rice plant the following spring. In so doing, it will bear hundreds of seed which, when sown the following year, could produce dozens of new lines of rice. By just adding a little help to this rice created by the insects, I get about a hundred new rice lines.

Since man is much better at creating new varieties than the insects, some college students who visited me thought that my discovery had little value. But I was greatly astonished by it all. This meant that what we think of as crop pests have all along been genetically upgrading rice, playing in effect a role in advancing the evolution of rice. Reflecting on the significance of this, I thought: "Why, that's incredible. Man should never have gone to all that trouble."

A hundred years ago, Darwin said that living things gradually evolve while adapting to nature. That has become the common understanding. But although theory has it that man arose from the monkey, the birds from the reptiles, and the giraffe from the horse, fossils of the "missing link" species have never been found. The lack of any evidence in support of evolution virtually negates Darwin's theory of evolution.

Reviving Ancestral Strains

Normal nonglutinous rice was developed from old varieties of glutinous rice. One can imagine from this that, along the way, a dizzying number of intermediate varieties arose, each slightly different from one another. Obviously, the genes of the past are present in old varieties. What this means is that ancestral varieties remain latent in the old genes. It occurred to me that by cross-fertilizing backwards in search of ancestral traits, I might be able not only to surmise what the ancestors of a cultivar were like, I might even be able to revive and resuscitate those ancestors.

After carefully organizing the hundred or so lines of rice that I have created recently and studying them from many different angles, I feel that I can say almost conclusively that glutinous and nonglutinous rice are sister races which arose from the same ancestors, and that upland rice and paddy rice belong to the same species. With further experimentation, I should also be able to determine what type of plant the ancestor of rice was.

It is just a thought, but instead of crossing plants of the same species, one could cross rice with the weeds crabgrass or couch grass, for example, to search for the common ancestral roots of both. Or one could perhaps try combining proso millet with barnyard millet, or creating a composite variety of foxtail millet and green foxtail. If successful crosses are obtained in this way, it should be possible to come up with new plants having the traits of the ancestors of rice or of these weeds. It even seems to me that by going a step further and crossing these new plants with each other, it might be possible to recreate the very first ancestral plants. I don't know, maybe this is just empty speculation.

Of course, all this will be difficult using only cross-fertilization techniques, but it should be possible with the various new methods available through biotechnology today. In any case, I think it possible to trace back and revive the "missing link" species of the past by making use of reverse cross-fertilization aimed at returning to the ancestral organisms from which arose differing species and genuses. In this way, we may come to have a better understanding of the route of evolution.

Problems with Darwin's Theory of Evolution ————————

But if such research were conducted, this could have serious drawbacks. Naturally, cross-bred varieties of millet and rice, both of the same family, would arise. In addition, we might get crosses between millet and eulalia or bamboo, and hybrids of bamboo and plum, or of plum and pine. Of course, the biologists would be delighted because this would elucidate the course of plant evolution from dicotyledons to gymnosperms and angiosperms. Naturally, this could all go too far, in which case man would regret what he had done.

Here is what I think about evolution. Darwin's theory would lead one to believe that ever since living things arose on the earth they have striven to evolve into subsequent organisms. But look at it this way. Living things apparently arose from nonliving things, but strictly speaking there is no sharp dividing line between the living and the nonliving. If we were to trace the course of evolution by living things, then we would have to return all the way back to the birth of the universe.

Insofar as living things are one with cosmic nature, one cannot think of them separately. If the living possess life, then so do the nonliving. The source of cosmic life is most certainly harbored also within nonliving things. Biological evolution cannot be discussed without taking into consideration the colossal evolution of all things in the natural world that also harbor life. There is a world of difference between our extracting living things from nature and considering the evolution of these isolated organisms from a standpoint in conflict with nature, and our seeing the entire universe as a single entity and observing organisms within that entity. What is important is not the evolutionary course taken by individual organisms, but the course taken by the life of all things.

What is the total cosmic life force? This, of course, is not the same life as that seen by biologists. Scientists today say that the source of life lies in the nucleic acid DNA. But nucleic acids certainly cannot be considered the unique source and carrier that transmits genetic information for biological life. Even though nucleic acids are present in the cell chromosomes, do we know from what the nucleus arose, or how and why protoplasm arose? Or we could go in the opposite direction and ask what nucleic acids are and break these down ad infinitum. What this indicates is that scientists have no idea what the first cause of life is. Nor is there any hope that they will grasp this sometime in the future. Nucleic acids are not the cause or the source of life, but just a manifestation of the result. They are merely one mode of expressing life, nothing more. The root of life lies outside the scope of science; it is not something that biologists can study.

Scientists appear to be under the illusion that there is some ultimate worth to the information transmitter we know as life. But what is of importance to man is who used this transmitter, why he used it, and what type of information he transmitted with it. What this means is that the transmitter which biologists are looking at is really just a relay station for transmitted waves.

I believe that the first cause of cosmic life is not to be found in the genetic material present within the cells of living things, but exists in a higher plane of reality. Of course, this cannot be seen or named. Perhaps, for lack of any better way of putting it, it could be called the "Holy Spirit" which continues to live within nature. Ultimately, this could be called "God."

This source of life that I take to be the first cause permeates nonliving and living things alike. I believe that, depending on the time and circumstances, it takes on a myriad different forms. But with his limited field of vision, man persists in labeling this evolution or regression.

The story of creation is not an ancient fable or hypothesis. I believe that what is happening *now* is the Creation. Or is it perhaps that I don't know how the word is used?

In any case, the animus in living things, which draws from the same well as the life harbored in all things, is a continuum of one and the same life. What appear to be different species are not that at all. Organisms intermediate to these surely existed in an unbroken chain.

I think that the lack of a fossil record of these intermediate species can be explained as follows. When two different organisms cross, this creates tens and hundreds of offspring. These offspring are not identical, each differing slightly from one another, but because they have continuous traits they form a discontinuous continuity. This is not to say that all shall thrive on the earth. Most of the intermediate species disappear; only organisms at either extreme or which are special in some way survive.

The reasons why most intermediate species disappear vary. I imagine that many species disappear before having had a chance to establish themselves, while others that have established themselves later fall out of sorts with the environment. In some cases, a lack of fertility may be the cause, while in others, offspring are conceived but fail to thrive. So, in what can be thought of as an accidental process, only a small number of the great body of potential species have managed to survive and establish themselves on the face of the earth. One could put this somewhat differently by saying that only a few intermediate species remained afloat, with the great majority disappearing under the surface of the sea. This could be called the "intermediate species sink-or-float hypothesis." In other words, all organisms are continuous on the ocean floor, but these are not visible to man, who sees only islands floating above the waters.

Another analogy that could be used is that of a heavy snowfall which buries all but the largest rocks and trees. The grasses and bushes out of sight beneath the snow never come within the purview of the scientist's research.

Darwin's theory, then, never really amounted to more than a temporary hypothesis. Although one must admit that this has done a lot to clarify the biogenetic lineage that shows what sort of organisms arose and vanished from the distant past to the present, the theoretical analysis is a biased West European view that misreads the true state of the natural world.

Arguments that posit concepts such as natural selection through adaptation, dominance of the strong over the weak, and survival of the fittest—which form the framework of Darwin's thought—as being the true image of nature are mistaken. To the human observer, the lion appears stronger than the rabbit, but nature most likely sees no such difference in strength. When acacia seeds fall thickly to the ground, which seeds will survive? The intellect cannot tell whether it is the organism that selects nature or nature that winnows the organisms.

The question of superiority and inferiority, of suitability and unsuitability, never existed in the natural world. Darwin's theory, which misjudged nature, has dealt the order of human society a blow of incalculable force over the past hundred years. In his myopic relativistic view, man discerns superiority and inferiority and observes the process of natural selection, accepting these as facts of life in the natural world. But these "facts" are based on standards set arbitrarily by man.

From the vantage point of nature, which transcends the relative world, nothing is in conflict. No distinctions are made, such as between insect pests and beneficial insects. The words fit and unfit, strong and weak, do not exist. Superior genes and inferior heredity are just near-sighted conclusions arrived at through the relativistic views of man.

There is no large or small, inferior or superior, in nature. Organisms are always equal; they live and change together in a constant flux, that is all. Organisms in the natural world know neither life nor death, progress nor retreat. Naturally, it is nonsense to think that superior adaptability provides the capacity for survival. This is nothing other than a profanity against nature.

Nature is always free of intents, actions, and plans; it has no secret purposes, no ulterior motives. The will of God cannot be comprehended within the limits of human knowledge. Hence man should never have subjected nature to scientific analysis. All the phenomena perceived by man are results; he is unable to grasp the true causes. That is why, although it may be permissible for scientists to report natural events as they occur, they have no right to comment or criticize.

Scientists Abuse Their Authority

Many people believe that the role of scientists is to develop theories that explain natural phenomena and to uncover basic truths about the natural world. But this is all wrong for, as long as he fails to grasp the first cause or purpose of nature, man cannot know the meaning, aim, or direction of life. So what man has done is to blindly set goals for himself based on standards established through the intellect and to start creating organisms and artificial man.

Even though the doctor can protect life and play a part in the birth of a child, he cannot bear responsibility for the life and death of that person. Even though he may succeed in creating test tube babies, he will be unable to hold responsibility for the entire life of that child from just a medical standpoint.

The medical goal may be to create genetically superior children, but who is going to decide what is superior? Only nature itself knows whether the rabbit is weaker or stronger than the lion. It is up to nature to decide whether an inferior gene should be eliminated from the earth or preserved. Although we may succeed in creating artificial man, we will not be able to predict the future of these creations of ours.

I would like to caution people that human actions and knowledge are effective only within a narrow range and produce, moreover, anti-natural effects. Regardless of how superior and seemingly natural the organisms that man may create, these can only be anti-natural progeny. As a result, the whole goes beyond the question of superiority and inferiority, and can lead only to kharmic tragedy. Anti-natural things are invariably incomplete, and will either meet with failure or proceed along a lonely path as the anti-natural offspring of man.

Of course, we have constant assurances that every care will be taken from the scientific, ethical, and religious standpoints, but this is not humanly possible. Suppose for a moment that a test tube baby rebels and pushes the nuclear button. Who then will take responsibility for the consequences? Just thinking of this is enough to give one a fright. The thought is even more alarming when one considers the thin line that separates the genes of a Nobel prize winner from those of a madman.

Most important of all is the realization of being man-made that an artificial person must carry with him throughout life. It is impossible to imagine the effect such a burden would have on that child. He will live a tragic existence who has been robbed from the moment of conception of the true freedom enjoyed by a child of nature. When this happens, wisdom and folly, superiority and inferiority, will no longer be the issue. The problem that will have to be resolved will be who will bear responsibility. The distinction between being born naturally and being born unnaturally appears to be very minor, so should something bad come of it, man and the physician will shrug and say, "But all I did was lend a hand." Artificial humans are unable even to accept responsibility for themselves. They will experience the misery of unimaginable isolation.

What does man depend on to live? I would say that it is the conviction he has deep down that he was born in nature and is sustained by nature. But the artificial person must carry throughout his entire life the unhappy knowledge that he will never have the peace of mind which comes from knowing that he was created and nurtured by the hand of God. No matter how noble the character of the scientist or physician, he cannot serve as a surrogate parent or a deputy for God.

Today, even in Japan, the ban has been lifted on recombinant DNA experiments on all plants and animals, with the exception of the great apes and man. It is the start of a dread age, but no one has raised a voice of dissent. This, most assuredly, is the seed of a new tragedy for the human race. Unnatural organisms are inevitably transgressors of God's laws.

Pine Blight:
A Case Study of Nature Under Attack

Over the past ten years or so, pine rot in Japan has spread throughout the country. At one time, long ago, there were people who held the view that the soil becomes depleted when red pines grow to their full size. Today, some scholars are even explaining away the rash of pine rot we are seeing as a case of plant succession by saying that the time has come for the pines to die off and be replaced by deciduous trees. But can we really rest at ease with these explanations?

It would certainly be nice if everything were perfectly normal and this pine rot could be accounted for as a natural process of succession. However, I feel certain that this rapidly spreading phenomenon we are seeing in Japan is not the result of damage by a single pest. In fact, I doubt that it can be characterized as anything other than a total aberration of nature.

I believe that nature today has become deranged on a global scale, that major disturbances have arisen in the soil microbe community. I think that the perishing of the *matsutake* mushroom symbiotic with the red pine has led in some way to physiological irregularities in the pines, making them susceptible to pests and bringing about their rapid destruction.

Eight or nine years ago, finding it hard to bear the strange sight of the large red pines near my citrus groves rapidly dying one after another, I brought out an old microscope and called upon my rusty skills. I knew full well that this was just the headstrong audacity of an old man, but I set up a small makeshift lab in one of my orchard huts and gathered together some basic supplies such as test tubes and petri dishes. I used a basket steamer as my sterilizer and a *kotatsu* heater as my room heater. I isolated pathogens and prepared pure cultures near the hearth. That is the sort of research setup I had.

For three years, I got up early before the sun rose and sat looking through that microscope. Sometimes I'd eat lunch, sometimes I wouldn't. In the evening, I returned home, went to bed early, and woke up again the next day at three in the morning. I really put myself through the mill, what with pouring through the literature and all. When I think back on it now, I sometimes wonder why in the world I did all this.

I suppose that the way others saw it, all this research didn't square well with my rejection of science. Youngsters at the farm would complain, as did one girl I can recall, that they had come to learn about natural farming, not to help me cut down pines and dig up roots. There wasn't much I could say to that, except that one ought to do what one can, even if that meant just trying to understand why those pines were dying.

There are limits, however, to the studies one can do in an orchard hut. But I did present a report of my findings at a symposium at Tsukuba University in the hope that this would provide some useful input on the problem. Here is a very brief summary of my observations and conclusions from a series of experiments and studies I ran over a three-year period. A more detailed account appeared in the June 1981 issue of *Atarashiki Sekai e*, a publication put out by the International Macrobiotics Center (Nippon CI) in Tokyo.

Pine Rot: Portent of Desertification ─────────

Pine rot, which was formerly thought to be caused by pine weevil infestation, is today believed to be the result of attack by pine wood nematodes at the twigs and branches of the tree. But the fact that this is preceded by considerable decay of the underground roots of the tree deserves close attention.

In my research, I found that the roots of pines are infested with a putrefactive mold which I will tentatively call *kurosen-kin* ("black bristle mold"), since I have yet to identify it. These bacteria completely destroy *matsutake*, a mycorrhizal fungus that is normally symbiotic with the red pine, causing the tree rootlets to turn black and decay. This is followed by the entry of a black mold that damages the larger roots, greatly weakening the pine.

Following decay of the roots, several new types of pathogen enter the tree along the trunk and branches. Most of these pine wood-decaying molds are not native to Japan, appearing instead to have entered the country on imported lumber. Since I have not studied the matter it is hard to say for sure, but several of these organisms seem to be connected in some important way with pine rot. Incidentally, during my visit to the U.S. in 1979, I was able to ascertain the presence there of both these organisms and the pine wood nematodes.

At about the time that the abnormal condition of the tree can be detected externally from the presence of pine weevils and long-horned beetles, these pests begin to lay eggs and various other putrefactive molds enter through the wounds made by the beetles in the bark. Pine wood nematodes also enter the tree at this time to feed on these molds, and multiply explosively. I believe that these, together with the invading molds, destroy the resin ducts within the tree, creating a severe wilting state which, during a hot, dry summer, almost immediately kills the tree.

At least two or three years before the pine wood nematodes invade the tree, the roots have already begun to decay, creating physiological disorders and reducing the amount of resin exuded by the tree. However, people do not notice the abnormal condition of the tree even at this time because the leaves remain green and there are few external signs of disease.

The reasons why the molds involved in pine rot multiply so rapidly may be the low resistance of Japanese pines; the fact that the red pine is a mycorrhizal plant, and mycorrhiza are highly sensitive to environmental changes and pollutants; and the high resistance of the pathogens to air pollution and pesticides. Another way of looking at this is that air pollution and pesticides have helped to provoke pine rot, in which case some radical and far-reaching measures to control these influences must be considered.

On the basis of my observations, I have some suggestions on how this problem could perhaps be brought under control. One step would be to raise the soil pH (which I found to be very acidic—about 3.2–4.8—in affected areas) to about 5 by administering lime or wood ash to the soil. A second step would be to pour large quantities of soil fungicide such as a mixture of Orthocide powder in water, or to spray the powder in stands of red pine shortly before heavy rains. Once this agent has disappeared from the soil, large amounts of *matsutake* spores grown in pure cultures should be inoculated into the roots to stimulate the growth of new mycorrhiza.

I think that one can regard the phenomenon of pine rot as an early sign of desertification brought about by a destruction in the balance between plants, animals, and microorganisms in Japan. I hope, therefore, that, along with immediate emergency measures over the most seriously affected area, drastic and far-reaching measures will be established to control these pests. Unless this is done, judging from the degree of root decay that I observed, Japan's pines west of Tokyo will be almost entirely wiped out within perhaps five years, starting the wholesale destruction of Japan's environment.

Several years ago, I visited America in July and August, during which time I saw for myself the horrible extent of pine rot over there. Speculation over its cause differs from that in Japan, where scientists persist, for whatever reason, in pointing to the pine wood nematode. In America, the cause is thought to be drought and air pollution by jet planes. I do hope that more serious efforts are made both in America and Japan to determine the cause of this disease and to find some effective way of controlling it.

The Natural Environment Must Be Preserved ─────────

It is only fitting that the first cataclysmic change to take place in the global microbial world occurred in a mycorrhizal fungus, which forms, after all, a highly advanced biological community in which many microorganisms are concentrated and organically interconnected. In a way, such a change arose where one would have expected it to. The red pine was able to become the strongest plant only because it was strongly protected by the *matsutake* fungus.

When humanity crumbles the first places to go under will probably be the great cities with their highly developed and congested societies. Pines protected by mycorrhizal fungi are incredibly strong plants capable of growing even in deserts and on sandy beaches, but without these symbiotic fungi they are weak and delicate. I feel that there exists a danger that when exogenous mycorrhizal plants such as the pines perish, this will spread next to endogenous mycorrhizal plants, to the Japanese cypress and cedar, to the zelkova, to the fruit trees, and perhaps even to rice. I'd be only too happy if all my concern was just needless worrying, but in Tokyo Prefecture, not a single red pine or Japanese cedar remains standing any longer. The next to go will probably be the Japanese black pine, followed by the zelkova. In Kyoto, the situation has deteriorated to such an extent that I believe this year may be the last chance we have for controlling the disease among pines at the shrines and temples there. In this sense, the question of whether we can protect the pine or not really boils down to whether we are able to protect Japan's natural environment. The Japanese as a people cannot survive the loss of Japan's natural environment.

Artificial Cultivation of the *Matsutake* ─────────

I would like to disclose here for the first time my dream of artificially cultivating the *matsutake* mushroom (scientific name: *Tricholoma matsutake*). In the course of my research on pine rot, I arrived at the conclusion that the key to restoring the red pine was to restore the *matsutake*.

One day, quite by chance, I stumbled upon a successful way to grow this fungus artificially while running pure cultures of the spawn. I was flame-sterilizing my materials in the hearth fire, as I have found this procedure to be more reliable than alcohol sterilization for pure cultures of wood pathogens and *matsutake*. Since I was trying to isolate mycelia from the caps of *matsutake* mushrooms, I was cooking the *matsutake* instead of sterilizing them. The *matsutake* is a delicacy in Japan so I dipped the cooked mushrooms left over in soy sauce and ate them.

I had been preparing different culture media using whatever materials were close at hand in my improvised lab in an attempt to determine what would work for *matsutake* spawn. When I dipped the cooked *matsutake* in soy sauce and tasted it, I found it delicious. It occurred to me all of a sudden that what man thought tasty, the *matsutake* might like as well. So I tried culturing the mycelia using a culture media containing soy sauce.

The fact is that even in the best culture media known for the *matsutake* up until then, the growth of this fungus was too slow. Moreover, the chemicals used in those media were not readily available to me in my remote location. The first successful test tube cultivation of the *matsutake* was done more than forty years ago, but this fungus has yet to be successfully cultivated in glass bottles despite the advanced pure culture methods available today, which says something about the difficulty of growing it. In fact, to a man, *matsutake* researcher all lament that no microorganism is as difficult to cultivate as *matsutake* spawn.

So my chances for success were very slim indeed. Only, instead of thinking of the *matsutake* as just a fungus, I tried thinking of it as an advanced form of life on a par with man. When I removed sections of mycelium from the caps and stems, I couldn't help feeling that this *matsutake* was in fact deformed, that it was a hard-to-please riddle of a microbe vastly different from the pathogenic fungi I had isolated from the pine roots.

In any case, as I was cultivating the pine rot pathogens, I started wondering if I might not be able to come up with a method for culturing the *matsutake* in bottles. I tried just about everything I could think of, and found that what the *matsutake* likes most of all is *chawan-mushi.**

To the average housewife in the kitchen, growing a pure culture of *matsutake* is out of the question. But when one prepares a delicious *chawan-mushi* and places a *matsutake* cap on top of this, the spores that fall from the cap grow well; in a half-year, the container becomes filled with white mycelia. If this is kept at a temperature of about 61 degrees Fahrenheit, little *matsutake* come popping up one after another.

The first time I saw four little *matsutake* come up in a one-liter bottle, I was surprised and delighted. I immediately took it outside in the sun to get a good photo of it. I got my picture all right, but I had forgotten that *matsutake* mycelia are very sensitive to high temperatures and will die in about an hour at 86 degrees. After all the trouble I'd gone through to grow them, the mycelia were done in by the heat. That shows just how hard it is to please the *matsutake*. But I had found that, with the proper technique and care, it is indeed possible to grow this finicky fungus.

I have called this culture medium a "natural medium." Even though the first experiments were successful, this is still a long way from being ready for practical application. I will disclose the full details of my method once I have had a chance to improve it some more. I am hoping also that the wait will arouse everyone's expectations. The day may be near when people will be able to grow their own miniature *matsutake* mushrooms in the kitchen.

---

* Japanese egg pudding.

Although I am urging people to try something new, my wish here is that everyone realize that the *matsutake* is a forest sprite that protects Japan's mountains and forests. It is not something that can or should be privately owned.

The artificial cultivation of forest *matsutake* in a bottle is the kind of thing that scientists dream of, but the moment I realized that this could lead to locking up this forest sprite in a bottle, my enthusiasm as a scientist suddenly cooled. That is only natural I suppose. Since then, my microscope has stayed in the closet.

Come to think of it, I accomplished nothing at all during those three years of intent research. This is as it should have been.

My true desire is to leave the secrets of the mountains as secrets. We must not create a spurious nature.

Keeping Out the Medfly

Fruit from Australia ────────────────

Given my experience long ago working in the plant inspection
division at the Yokohama Customs Bureau, I was surprised
by news of the outbreak of the Mediterranean fruit fly in
California. Perhaps it was uncalled-for meddling on my part,
but after meeting officials at Yokohama Customs and the
Agricultural Ministry, I wrote the following piece for the
Yomiuri Shimbun, one of Japan's major general-circulation
dailies. This ran on September 17, 1980.

*   *   *

Keeping the Medfly Out of Japan ──────────────

Will the Line of Defense Hold?:   Over the past several
years, I have spent what time I had left over from my
farmwork in studying the problem of pine rot. Since my
investigations led me to believe that this disease may have
arisen from microbes brought into the country on imported lumber, I began to wonder how custom inspections are done on lumber imported into Japan. Just as I
was turning this over in my mind, I came across a small
article in the newspaper on the outbreak of the Mediterranean fruit fly in California.

Medfly Resistant Even to Deadly Pesticides: This was
one of those cases in which big news are hidden under
a small headline. The article may not have been noticed
by most people, but it was very important news for
Japanese farmers. Smaller than a housefly, the medfly is

a beautiful fruit fly only five millimeters long. The larvae, which bore into and consume fruit, may well be the world's worst insect pest. If it were to enter Japan and multiply from April through October, this would be enough time for several generations to emerge and sweep throughout Japan. The medfly feeds on a great variety of crops. In Japan, it would attack some 45 families of 143 plants, and might very well deal a lethal blow to most Japanese fruits—including the peach, plum, apricot, and apple—and fruit vegetables such as the tomato. This is an extremely bad pest that cannot be controlled by spraying deadly pesticides or the release of sterile mates.

Too Few Researchers: The medfly is native to the Mediterranean coast, but has spread to Africa and South America and is feared in all subtropical regions. Naturally, all fruit imports into Japan from countries with the medfly are banned. The major aim of Japan's Plant Quarantine Law, which was established in its original form in 1914, has been to prevent the entry of fruit flies from low-latitude countries and the codling moth from countries at colder latitudes. Fortunately, for sixty years, plant protection efforts at customs have succeeded in keeping these out of the country. However, the outbreak of the medfly in California could conceivably escalate into import prohibitions on all orange and other fresh fruit from America. This is frightening news for farmers of both countries.

As a former plant inspection official at customs and as a fruit grower today, I was deeply concerned about this development so I hurried over to Yokohama Customs and the Plant Protection Division of the Ministry of Agriculture, Forestry and Fisheries to find out what the situation was and how the government intended to deal with it. One customs official that I talked to told me that Yokohama Customs had received a warning to be on immediate alert from the American Embassy and was conducting rigorous inspections of incoming produce. It was clear to me that the mood at the plant inspection division was tense. While there, I found most surprising the fact that, although the number of inspectors had increased a hundredfold with the sharp rise in plant imports and exports, the number of researchers, who are vital to an effective inspection program, was the same as it had been forty years ago; namely ten each in the insect and pathogen sections. I could well imagine the fits of anxiety the medfly scare was causing in the manager of the insect section who had to handle this problem all by himself.

To prevent the entry into the U.S. of the medfly from the Caribbean Sea area, America has laid out a broad, million-acre "line of defense" along the Mexican and Guatemalan borders over which two helicopters constantly spray organochlorine agents. Other extermination strategies are conducted here as well. Naturally, the logistics and expenses involved are enormous. The U.S. government has announced that the medfly: 1) is believed to have been brought in from the Hawaiian Islands by a traveler, 2) broke out in private orchards in Los Angeles and Santa Clara Counties, but has not arisen in commercial orchards, and 3) is under complete control so there is no need for concern over fruit headed for Japan.

But, if I may venture a personal opinion, I believe that there is reason to fear that the infestation will break through the Mexican "line of defense." Moreover, if the fly did indeed break out in urban areas and private orchards, then one cannot assert with any assurance that the medfly has not reached the commercial orange groves.

The destruction that can be caused by imported pests is clear from past examples. The three major citrus pests in Japan—the arrowhead scale, the cotton-cushion scale, and the red wax scale—were all imported around the turn of the century. Ever since then, orchards have had to be sprayed with pesticides. A couple of decades later, the wooly apple aphid was brought into the country. From that time on, apple production in the Kansai district of Honshu has been very difficult, which is why apples today are grown primarily in northern Honshu where aphid infestation is slight. And the fuss and holler that went up with the outbreak of the fall webworm on the trees lining city roads is still a recent memory.

When pests of foreign origin enter Japan, this causes lasting harm to the farmer. If Japan tries to make the best of a bad situation while awaiting further details from America, it will be too late. And there is little doubt that dragging out the problem with stopgap measures will leave eternal regrets. Japan and America must not hesitate to do everything in their power to arrive at a full understanding of the threat posed by the medfly and to reinforce their defenses.

Is Arrival of the Medfly Just a Matter of Time?: To me, it appears as if the epidemic of pine rot has triggered an avalanche of environmental destruction in Japan. Nature is in the grip of annihilation. Given that this country is totally absorbed in its efforts to expand food imports and exports as part of the doctrine of an international division of labor, I suspect that it may be only a matter of time before the fruit fly arrives in Japan. If even a single fruit fly finds its way into the country, Japan's farmers will enter an even longer and darker tunnel.

Needless to say, we can do no more than wait for the entomologists and agricultural scientists to become aroused to action against this beautiful little fly. I would call especially for careful reflection on the fact that a lot hinges on the success or failure of plant protection in

Japan, and on the fortunes of Japanese agriculture. Along with an emergency program of basic research on controlling infestation once the medfly reaches Japan, I feel that, rather than fighting the pest at the water's edge, it would be wiser for Japan also to set up a major "line of defense" of the sort that the U.S. government has established in cooperation with the Mexican government.

Fruit Flies from Australia

The medfly became a major issue for a year or two after that, but I never, in my wildest dreams, suspected that a fruit fly would make its way to my home in a remote corner of Shikoku. It was a trick of fate that unexpectedly brought several fruit containing flies to my door and sent me into a mad flurry of activity for ten days. On April 21, 1984, all the major national papers carried news of the arrival in Japan of one of the world's most feared fruit flies. The entry of a single fly brought with it the danger of a sweeping devastation of Japanese agriculture. The following article appeared in the *Asahi Shimbun*.

*   *   *

Fruit Fly Maggots Discovered in Shikoku

The maggots of a fruit fly which is feared could cause heavy crop damage were discovered in fruit received as a gift by a farmer in Iyo-shi, Ehime Prefecture from an Australian visitor. The maggots were brought to the local branch of the Agricultural Ministry's Kobe Plant Protection Center on April 19. Entomologists at the Center identified the maggots as being those of the Queensland fruit fly. Up until now, foreign fruit flies have been stopped at the plant quarantine stations at international airports and ocean ports designated for plant import and export. This is the first time that any have been discovered after being brought into the country. The Kobe Center dispatched two senior inspectors to Iyo-shi for a follow-up investigation. With the cooperation of local officials, they set up about fifty monitor traps in fruit trees and huts near where the Australian had spent the night. The center has also launched a full-out effort to determine the route of entry of the fruit.

The farmer who discovered the maggots was Masanobu Fukuoka, who practices natural farming in Iyo. An Australian acquaintance visited Fukuoka on April 17, stayed overnight in his hilltop orchard, and the next day gave Fukuoka three guavas as a present. When he later cut open the fruit, Fukuoka found twenty to thirty small white worms about one to two millimeters long in each of the fruit. Having worked at one time as a plant inspector at Yokohama Customs, Fukuoka immediately realized that these were fruit fly maggots. The Australian told Fukuoka that he had only brought three fruit with him, all of them picked near his home. Concerned nonetheless that he may have brought in other fruit as well, the Agricultural Ministry questioned Fukuoka on the whereabouts of the visitor, but the Australian had said only that he was heading for Kyūshū.

According to officials at the Yokohama Plant Protection Station, Australia is home to the Mediterranean fruit fly as well as the Queensland fruit fly. Both pests cause enormous damage there to fruit trees and other crops. The medfly was discovered in California in the summer of 1980 and rapidly became a major problem, taking two years to bring under control.

The import of fruits into Japan is strictly controlled under the Plant Communicable Disease Control Law. All fruits from Australia are banned, with the exception of pineapples, coconuts, and unripened bananas. Importation is allowed only at international airports such as Haneda and Narita and at eighty designated seaports throughout the country, and only after passing through a quarantine based on formal notification of the shipment. But nothing can be done when a traveler brings in unreported fruit as in this case.

If one of these fruit flies were to break out in Japan, not only could this result in widespread damage to the nation's fruit crops, it could halt the export of fruits from Japan under the International Plant Protection Convention.

Plant protection officials in Ehime Prefecture intend to continue periodic investigations, but because the maggots discovered had not yet emerged and temperatures remain low, there seems little danger of the fruit fly spreading.

First Appearance of the Fly in Japan: The Queensland fruit fly, the adult form of which has a length of six millimeters, occurs naturally in Australia, Papua, New Caledonia, and other parts of Oceania. It feeds on most other fruits and causes up to one hundred percent damage. This is why Japan bans the import of all fruits from Australia except pineapples and unripened bananas, which the fruit fly does not attack.

In the period of a single year (1983), the Oriental fruit fly, the melon fly, and the medfly, all of which are major fruit and vegetable pests, were detected at Japanese airports and seaports 255 times from gifts of fruit and other produce brought in by people returning to Japan. In each case, entry was halted at the port of entry. The present case is the first known instance of a major fruit fly pest actually being brought into the country.

*   *   *

Needless to say, I was astonished to find small white maggots in fruit received as a gift the day before from this young traveler who had come to invite me to visit Australia. I immediately sent the fruit to the plant inspection divisions at customs in Kobe and Yokohama. A few hours later, I received a call confirming that these were indeed fruit flies. When the news that a fruit fly had entered Japan was broadcast over the national networks the next day, it created an uproar. Tense directives went out from the Agriculture Ministry's Plant Protection Division to all parts of the country and two head inspectors from Kobe Customs arrived immediately at my farm with a top local official. Over the next ten days, they worked night and day trying to track down my visitor and combing my orchard for signs of the fruit fly.

The Australian was traveling about in a small camping car, cooking for himself. These officials were in a frenzy wondering whether he had brought other fruit with him and was throwing away the debris after eating them. The danger was that, given another ten days, the maggots would emerge as adults. Once they had taken wing, it would be too late to do anything.

For several days, I too stayed by the phone, Whenever a little information on him came in, I got swept up in all the commotion. On the tenth day, just about the time the maggots would emerge, he presented himself at the customs station in Shimonoseki. At last we were able to get some details and things finally settled down.

But the problem has only just begun. This incident clearly shows that Japan's "water-front" defense against entry of the fruit fly is inadequate. The question now is what to do next.

What I'd like to point out here is that it is extremely difficult to completely treat fruit that is brought into Japan. The chances of detecting the pests in every instance are very small. In the space of a month, a single fruit fly can become a thousand flies. In two months, these can multiply to ten million, and in three months, to an essentially infinite number.

Japan—A Country Without Fruit Flies ——————————

After the hullabaloo over fruit flies had died down, I got a jolt when one foreigner said to me, "The Japanese are fooling themselves if they think they can both protect and utilize nature."

People in other countries were startled at all the fuss made by the Japanese over a single fruit fly. This surprise came from learning that Japan is a country whose fruits are free of all fruit flies. They were envious, saying what a wonderful place Japan is and how fortunate its farmers. To people in Europe, Africa, Australia, and South America, it seems almost a miracle that there is a country on earth without fruit flies. They say that it is normal to find worms in fruit. Some just cut out this part of the fruit and throw it away while others eat the fruit, worm and all, figuring that this too is nutrition.

Even in Australia, which seems a land amply blessed by nature, fruit growers are legally required to spray pesticides. One visitor from Australia told me that this creates a dependence on pesticides and lamented such spraying as the start of the destruction of nature. "Why is it," he said, "that in a country without fruit flies, the farmers spray pesticides and destroy nature? Pesticide application isn't even required in Japan, is it? And why, when all fruit imports from Australia are banned, does Japan not also ban imports from California, with its outbreak of the medfly. It's just not fair." He also called the Japanese unreasonably optimistic because he saw no evidence of their taking a more rigorous attitude toward nature and national legislation, or of the Japanese farmer taking greater personal responsibility.

Another foreigner even questioned the degree to which Japan's farmers are aware of, and grateful for, the efforts and diligence of the plant inspectors at customs offices around the country who for sixty years have managed to prevent the entry of a single fruit fly into Japan.

What indeed will become of plant protection efforts in the future? The question is one that causes no small anxiety.

A village in the Somalian savanna.

Sowing seed in Somalia.

Visitors gathered in the orchard on Thomas Nelissen's farm.

Talking of God beneath the towering trees at French Meadows.

Scattering seed along the roadside at the Oregon Pass.

Giving a lecture at Davis, California
on how to set up a home garden.

A bird stops to feed out of my hand, in a forest near Olympia.

The author speaking at a church in Berkeley.

In Ashland, Oregon, we formed a group to sow seeds
by plane and revegetate these parched hills.

Exporting feed for livestock depletes the land.

An orchard in Eugene, Oregon. The surrounding land is semiarid.

A ceremony at the Lundberg Farm, to organize
local farmers for growing natural rice.

**The three Don Quixotes: Bill Morrison (left),
the author (center), and Wes Jackson (right).**

Spring at it height on my natural farm.

The author, dressed in a *yukata*, relaxes with his camera.

**Surrounded by daikon and rape flowers
on my farm in Shikoku.**

4

Natural Farming:
A Personal Testimony

The Principles and Practice
of Natural Farming

Here is the text of a speech outlining my experiences and approach to natural farming that I gave before members of the Sekai Kyūsei Kyō (The Religion for the Salvation of the World) in January 1975.

*   *   *

I graduated from Gifu Agricultural College and at the age of twenty-five joined the plant inspection division at the Yokohama Customs Bureau. There I did research in plant pathology and worked as a plant customs inspector for a while. I spent countless hours looking through the eyepiece of my microscope. As I did so, I noticed that the tiny worlds of the fungi and bacteria have something in common with the vast universe of heavenly bodies. There are males and there are females in those little fungi too. At the time, I was working on crossing molds. Molds bear a close resemblance to man not only in their shape, but in everything that they do. While I was pursuing these thoughts, filled with doubts and wonder, I fell ill. Then one day, prompted by a chance incident, I underwent what I suppose you could call a conversion of faith. It was a turning point. I'm not going to get into that here, but I had the feeling then that science was some sort of outrageous monster.

Sensing that everything is utterly meaningless, I quit my post at customs and headed back to Shikoku. I didn't head straight back, though. I traveled around a good bit, during which time I conceived the idea of

natural farming. On my return to Shikoku, I retreated
to my father's orchard to try this out. This was during
the early years of the war. As the war escalated, a life of
leisurely isolation in the hills became impossible, so I
joined the agricultural testing station in neighboring
Kōchi Prefecture, where I was placed in charge of insect
damage and worked through to the end of the war. While
at the Kōchi testing center, I did scientific research on
farming methods and ran around providing guidance and
instruction to the local farmers on growing rice and barley and on encouraging seed germination. Our goal at
the time was to maximize food production for the war
effort. At the same time, however, I had this idea of
natural farming in the back of my head. So along with
the scientific research I was doing, I also did some research of my own on natural farming. When the war
ended, I was free to go at last and become a farmer as I
had desired. I wasted no time in putting my ideas into
practice.

So I was still a youth when I happened upon the idea
of "do-nothing" farming. But although I knew that such
a way existed, I had no idea at first how to carry it out
in practice. I didn't know the methods. For thirty years
since then I've farmed in search of those methods.
Eventually, I came to have some idea of what these are.

Is "Do-Nothing" Farming Possible? ───────

What I do for a living today is run a citrus orchard on
a hilltop overlooking the Inland Sea. In addition to my
7–8 acres of mandarin grove, I also have an acre and a
half of paddy field on which I grow rice and barley.
Over the past several years, a constantly changing group
of about six to ten young people have lived in huts in
the orchard, helping me out with the field work. It's not
really a community as such. I've been calling it the "natu-

ral farm," but I don't even have a sign up to that effect. Those who come there to live and work are generally people following a natural diet, members of the Yamagishi-kai—an organic farming association, Shintoists, Christians, and members of other religious bodies such as the Sekai Kyūsei Kyō and Ōmoto*. I get young people and hippies, gamblers and villains, college students and runaways. Everyone comes as they wish, stays as long as they care, and leaves when they are ready.

So what do I do? As I just said, I grow rice, wheat, and mandarin oranges, but I've placed the greatest importance on rice. I consider myself above all a rice farmer. My technique for growing rice requires that as little be done as possible.

It was inevitable, both theoretically and ideologically, that I come to the conclusion that a way of farming exists that requires nothing to be done. That's because the first principle of my system of thought is that we do not understand; it is not possible to know and understand. The second principle is that nothing, no matter what it is, has value in and of itself. And the third principle is that anything done with the human intellect is worthless; it serves no purpose. In a word, all is unnecessary. When I arrived at this conclusion, I lost a standard by which to judge what is true and what is false.

In the Buddha's Heart Sutra, it is written that "Form is emptiness and emptiness is form. All is nothing." The way I see it, if one accepts this verbatim, then it means that all effort is in vain. Buddha is saying that we are not alive, we are not growing, and we are not dead. Most people would react to this by thinking: "Oh, come on now. What nonsense. That flower over there on the table is alive. It's not wilted or dead, right?" But the Lord Buddha says that this is neither dead nor alive.

---

* Both new religions in Japan.

People think that they are more than flesh and blood; they believe they also have a soul. They understand the body by the beating of their heart. But the Buddha says that the soul itself is just something that arises from the body, and that neither exist; that the flesh and the soul are emptiness. He clearly and unequivocally affirms that this body which so clearly seems to exist does not. One wonders indeed how he can say this, but if we take what he says literally and accept that we can know nothing, then this means that there is nothing, that neither concrete phenomena nor mental images exist. No matter how hard I try to show this to others, it never gets through. But even though I was unable to get others to understand, I wanted to try to demonstrate it for myself. So I set out to confirm this while farming in the fields. I did not take up rice farming to speak to people about my ideas. Farming was not what I did at all. I tried to establish whether this idea that we have no need to do anything is valid or not. I wanted only to find out for myself whether the words of the Buddha in the Heart Sutra are really true or not. All I did was to try to determine whether rice can be grown without doing anything.

While I was at the agriculture testing station in Kōchi, I also tried the opposite approach: "What about doing this? What about doing that?" But this only makes people busier and makes things harder for the farmer, who ends up having to do everything. Preoccupied with the thought of having to do this and that, people are forever worrying. So what I did when I became a farmer following the war was to search for things that don't have to be done. I asked whether the fields really need to be plowed and turned, whether the rice farmer really has to transplant his seedlings, whether it is necessary in fact to spread fertilizer on the fields.

The Paradoxes of Scientific Farming ──────

From my experiences, I learned that it isn't necessary to do anything. Scientific truth appears to be truth, plain and simple; but it is only *scientific* truth, not *absolute* truth.

Things appear to have value, but invariably man has set up the conditions that give them their apparent value. For example, the reason a field must be plowed is that man has created the conditions that require this; he feels as a result that doing so is worthwhile. The farmer floods his rice paddy for six months and runs a tiller through it, turning the soil into something akin to adobe mud. The constant activity kills the microbes and drives the air from the earth. As the soil is kneaded, this breaks down the clumps of earth into smaller and smaller particles. Eventually, whether he likes it or not, the farmer must run his plow deep through the field to let air into the earth. He weeds and intertills, stirring the soil repeatedly. All that is extra work. Man creates the conditions that require him to plow each year.

What would have happened had the earth been left alone to begin with? Leaving a cultivated field untended is abandonment, but that is not what I'm talking of. If the land is not abandoned, but instead left in a natural state; if, as in the mountain forests, the earth is not plowed or tilled, but left to itself, in a few thousand years, a rich soil develops.

Man kills and destroys the soil. Then he packs it into these pots, takes it to the laboratory and runs tests on it. Based on the results of such tests, he concludes that by increasing the plowing depth from one inch to two inches, one can harvest an extra five bushels of grain, and that plowing to a depth of three inches will give even higher yields. Of course, all that makes perfect sense when you take dead soil and subject it to tests under unnatural conditions. So word gets out that the deeper one plows the higher one's yields. Well, the fact is that even if one deep-plows, one can get good yields. In China, for example, they plow the soil down to three inches and still get high yields. So there is no doubting that it can be done. But they only do so because they're operating under the assumption that this has to be done. Wouldn't it have been better had there been a way to grow those crops without plowing? The reason chemical fertilizers have to be applied is that the paddies are filled with water and the roots of the rice allowed to rot so that the plants are weak. With the roots half-rotted, the crop will be lost unless quick-acting chemical fertilizers are used. That and only that is why administering chemical fertilizers has a booster effect on the rice plants. When man establishes the conditions in a field that render chemical fertilizers necessary, plants rice in that field, then runs tests in which he fertilizes one section of the field and doesn't fertilize another, it stands to reason that the fertilized plot will produce larger plants. That's why people think rice can be grown in high yields with fertilizers.

But when pots are taken out into a field, filled with soil and brought back into the lab, that no longer is natural soil. That soil has died. And when the field is flooded with water and rice is planted there, just adding organic matter is not enough for the rice to take. You need some-thing that will act immediately. This is the same as a sick person who needs a special diet. Chemical fertilizers that serve as a special diet for weak rice are not necessary. If the soil were left intact and healthy rice grown, there would be no need for chemical fertilizers. Developments in chemical technology create earth that has to be plowed. And with the frail, leggy rice that grows as a result, pests break out unless the plants are sprayed with pesticides. So they are sprayed. All that was really necessary was to come up with some way of enriching the soil while leaving it in nature's hands. But instead of trying to breed healthy rice, scientists breed rice suited to these artificial and unnatural methods of cultivation. Far from upgrading rice, then, breeding programs have succeeded only in downgrading it. All that has been accomplished is the creation of weak rice under the prompting of consumer demands for "tasty" rice, the creation of a rice that must be sprayed with pesticide. From this we have concluded that spraying rice is beneficial. It is not that I enjoy the role of devil's advocate or seek intentionally to be contrary, but I've followed the reverse course of not doing this and not doing that. Merely leaving one's fields alone is not nature. While walking the fine line between nature and the laissez-faire approach, I have striven to this day to learn what the true form of nature is.

Naturally Farmed Rice

This here is rice I harvested last fall. I assure you that I wasn't particularly choosy about the specimen I brought, but as you can see this plant is entirely free of disease. There isn't a single spot of disease anywhere. No rice blast disease, no sclerotic disease, and no leaf blotch. All one can find are a few marks left by locusts. That plant over there is winter-sown rice, which I planted before New Year's. This other plant here is a sample of the same variety sown in June. The two are slightly different. One has 12–13 stalks and the other 17–18. That's how much these tillered. This rice was sown at an interval of about five inches. There is a good deal of variation in the number of grains on a head. The smallest is about 120 grains and the largest about 260–270 grains. The rice I grow generally produces up to 12 tillers with an average of 250 and a maximum of 300 or more grains per head. The number of heads is about the same as ordinary rice and the number of stalks seems about normal when looking at the plant, but when you do the calculations there is no mistaking what an average of 250 grains per head comes out to in terms of yield. Maybe I'm stretching things a wee bit, but based on the figures I get—and I've checked and rechecked these—this rice can give yields of up to 50 bushels per quarter-acre. This is a yield higher than that gotten through scientific agriculture. The top theoretical yield of rice grown under the open sun has been calculated at from 50 to 70 bushels, so my rice isn't far off.

Let me explain now how I grow this rice. For twenty-two or twenty-three years I haven't tilled my field, applied chemical fertilizers, or sprayed pesticides. And yet I reap this rice. What I have here is an old variety of glutinous rice. I'm not exactly sure how old, but I think it was grown in the Tokugawa Period (1600–1868). The reason I grew this was to determine the yields that could be obtained with an old variety of rice rather than the new high-yielding varieties that breeders have developed. I also wanted to see whether it was tasty or not. I had heard that the samurai used to eat glutinous brown rice while the farmers ate sweet potatoes and millet. Maybe the reason samurai were able to mount their steeds, don helmets and armor, and charge into battle brandishing their swords was that they ate this brown rice.

I also tried to determine how good this rice tasted. I found that it tasted pretty good eaten as brown rice, but wasn't very appetizing as white rice. When I fed it to the young people with strong stomachs living in my orchard, they thought it was the best rice they had eaten. This raises serious doubts as to the credibility of all those claims that new varieties of rice are better-tasting and higher-yielding. I tested these assertions and found them wanting.

Tell me now, everyone: What happens if you don't have to plow, if you don't have any need for tillers and

tractors, if you don't have to spread chemical fertilizers
and pesticides? The key sectors of Japan's chemical
industry will all collapse. No wonder then that no one
is willing to believe such a method of farming is possible.
Look at those university agricultural testing centers everywhere and what they're studying. All that new technology
is only hurting the farmer. The labor-saving methods
being developed by armies of researchers are squeezing
the farmer tighter and tighter. This is reducing the number of farmers. We're seeing a boom period in agricultural technology that is producing chemicals and fertilizers which are wrecking agriculture. The entrepreneurs
have found themselves a new way of making money.
We are seeing a golden age of farmer-less farming. Farming methods today have done nothing but torment and
squeeze the life out of the farmer.

The Pit of Knowledge

What sort of thing does a college do? Properly speaking,
the goal of a college should be to create people who are
not lost, to create sages—people without doubts or illusions. That is why students learn things at school. Long
ago, I think the purpose of study was to create individuals who have no need to know anything. But colleges
today are different. Everything has been broken down
into disciplines, discrete fields of specialized learning have
been created in which the more one studies, the less one
comes to know the world at large. People look at a single
drop of water or a single rice plant and, with this as the
focal point of their inquiry, come up with an interpretation. Research is fragmented. Thus, if there is a spot on
a leaf of the rice plant, this is examined by a plant
pathologist; if an insect is found, the entomologist takes
a look at it; fertilizers are studied by fertilizer specialists;
and crop cultivation is the domain of the plant husbandry

experts. Everything is broken up into discrete domains and studied. The point is, are people able in this way to know a single rice plant?

When you come right down to it, you have to admit that man is incapable of knowing what this thing we call nature is. I began a talk at a university once by asking the students whether man is capable of knowing nature. A third of the audience raised their hands in the affirmative, another third said that man cannot know nature, and the remaining third gave no response. Do the students who answered "Yes" really know what it means to understand? Man is incapable of analyzing nature. Those who respond that they understand do not understand. They don't understand what it means to understand.

If one were to understand, then things would become clear and one would be able to arrive naturally at a lucid explanation. Once we are able to look at the rice and understand it, then a discipline devoted to rice is no longer necessary. But universities today just study and study, tangling the object of investigation into more and more complicated knots.

Let us say that we have here a single rice plant. To study this plant, all we have to do is watch it closely. But one must not look at it with doubt. One must believe the rice and observe it by putting oneself in the place of that rice. All we have to do is grow the rice, but in order to look at it we break it down into its component parts—leaves, stems, and roots—and examine these. It is as if man is digging and enlarging an underground pit. All that we have to do is look at the rice growing here under the sun. But scientists examine it under the microscope; they take it back to their dark laboratories and study it there; or they examine rice planted in small pots. Then they ask the question, What is rice? Their studies tell them that plant pathology gives one answer and plant husbandry another. These results are added to the growing body of scholarly knowledge on rice. We say that we understand rice because plant pathologists investigate the pathology of rice; fertilizer experts look at rice from the vantage point of fertilizer science; and economists study the economics of rice cultivation. As this underground pit of knowledge is enlarged, the darkness of the unknown surrounding it grows. One bit of knowledge creates two new queries in our minds. Solving these gives rise to yet new questions. To say "I understand" is to not understand. At the universities, "I understand" means nothing more than to have broken what we are examining down and wandered into the obscure darkness of ignorance.

Ignorant Agriculture, Misguided Medicine ————

It is just like the colors black and white. One is defined in terms of the other; ask what white is and the answer will be that it is the opposite of black. And this is what people think it is to know. But they don't know that white flower or anything else. I'll bet that not one person here can look at that flower and say he "knows" it. Ask what a flower is and you'll find that no one really knows. No one—not a single human being—knows what that flower is saying, what sort of will it has, whether it is speaking to man, whether it is thinking of something or of nothing at all, what sort of existence it has with respect to man—and whether that is real or fictional.

Yet, in spite of this, we dissect and analyze it, calling it this or that type of flower. Once he knows the name of the flower, the botanist is satisfied with a botanical explanation. The photographer brings his camera and takes a photograph of the flower. If he takes a black-and-white shot, he is convinced that it is a black-and-white flower, so he calls it white or black. The fellow who takes color photographs merely believes that the colorful flowers he has captured on film are the real things. All he's done is to capture on film what he has seen with his own subjectivity. Instead of catching hold of the object photographed itself, he's just having fun photographing what he has interpreted within his own mind and is under the impression that he really does understand. This sort of thing has not brought us closer to understanding but has multiplied the doubts that lead us away from understanding. That is why illusion just deepens further and further. It is as if people had dug themselves a pit in the ground without knowing any better, lit the lamp of ignorance there, and were rejoicing at the establishment of some subterranean city in a bright and unfailing civilization.

So I think of colleges as places that create and dispense doubt. They generate an endless procession of doubts and riddles, making it necessary to set up new classrooms and laboratories for solving these and to add on more faculty members. The more progress that is made in the fields of learning, the larger the school grows, until eventually it reaches gargantuan proportions. Man has created enough universities to flood the world. These have grown and thrived to such an extent because the world we live in has become so complex and incomprehensible. The universities have thrived because our doubts and questions have multiplied. While humanity has gone further astray, the universities that were meant to correct our benighted state have just grown and grown.

The number of doctors is increasing and hospitals are thriving too. Physicians think that they are curing the ill. If you create people of poor constitution, then conduct research in pathology and continue to weaken the human body, the number of research topics will go right on increasing, spurring on the field of medicine and enriching the medical profession. Let people eat delicious foods or whatever they like. As long as there are dentists around, there's nothing to worry about. Since it's okay to let the children eat whatever sweets they wish, mother gives the kids lots of goodies—as much as they want. Then, when the cavities start showing up, she takes the kids down to see the dentist, who gives them fillings or even new teeth. Happy that the cavities have been "fixed," she goes on giving her kids candy, certain that the more sweets the children get, the better off they will be. That's why people feel so reassured as our hospitals get larger and new advances are made in medicine. We rejoice over our longer life spans and our greater happiness. All that has really happened is that today we've got lots of senile oldsters and fewer young people, but for some reason everyone is celebrating progress in medicine and the longer life expectancy of people. The fact of the matter is that as medicine becomes more advanced and the number of hospitals increases, the human body becomes frailer. The advance of the hospital is merely a barometer of the collapse of the human body.

Agricultural progress has really consisted of nothing more than resorting to passive rescue efforts through agriculture when the rice grew diseased and unhealthy and yields fell. Although natural farming was capable of growing and yielding eight and even twelve bushels per quarter-acre, man destroyed the soil and he destroyed the rice by breeding new, weaker varieties. So, of course, yields fell gradually. Mankind has put himself in the position where, unless further progress is made in agriculture, it will no longer be possible to harvest the yields of yesterday. Scientists think they've been conducting research to increase yields, but all they've actually been doing is to develop techniques for reducing declines in yield. In a sense, what agricultural science boils down to is this: First man creates unnatural and incomplete conditions. Then he develops fields of learning and technology to restore the productivity of the land. Finally, when these appear to achieve the desired aims, he brandishes them proudly for all to see.

Is Natural Farming Catching On? ────────────

Agricultural testing stations in Shikoku and parts of Honshū have begun studying my farming techniques. The conclusion reached so far is that direct-seeding, no-tillage rice/barley succession seems to be okay. The research station at Okayama recently published some articles that describe the successive cultivation of rice and barley by direct seeding. This is almost identical, word for word, to what I wrote some ten years ago. There is one important difference, however. The researchers add that yields might be improved somewhat if, in addition to these natural farming practices, a little pesticide were sprayed. They also mention that it might be convenient to use a bit of chemical fertilizer. So what they're trying to do is to clothe natural farming in the robes of chemical agriculture.

I pointed out ten years ago that transplanting rice seedlings is totally unnecessary; that regardless of what new developments arise, there is no point at all in transplanting. Lately, the way things look today, I've gotten to thinking that the amount of transplanting done may drop by half over the next few years. At the very least, I believe that there are some grounds for hoping that my method of direct-seeding will be adopted around the country. Granted, the prospects seem to have receded somewhat with the advent of mechanized rice planters, but I think this is only temporary. Farming equipment manufacturers are really raking in the profits right now, yet somehow I just don't think that's going to last very long. After all, no matter how you look at it, yields with direct seeding methods are better than when transplanting is used. Agricultural technicians are not ignorant of this fact. They know, but for one reason or another they're vacillating—perhaps they're just afraid to come out and admit it. In any case, I'm confident that the tide will eventually turn toward direct-seeded cultivation.

The question of whether to shallow-plow or to practice direct-seeded, no-till cultivation is a point of contention only among researchers. The problem is that research stations have no experience with long-term, direct-seeded, no-till cultivation. I suppose that, until they gather the data they need, temporary tests of direct-seeded cultivation with tillage are all right. But no-till is the easiest method of all and to one's advantage. This is why I believe that the day of direct-seeded, no-till cultivation will come. When this technique does begin to spread, that will be the time to bring natural farming out into the open. And I feel that the time is almost here. Why, last year and the year before, do you know what method of cultivation got the highest yields at the Ehime Prefectural Agriculture Testing Station? That's right, direct-seeded, no-till cultivation. In fact, that grain was winter-seeded. The seeds were sown in October and December. Both barley and rice were seeded together. This method was very similar to the method I first reported in 1961. It's taken them awhile, but a few researchers in Ehime and Aichi Prefecture have at last given it a try.

My Method of Natural Farming

In early October, what I do is take no more than maybe a pint of clover seed and, pinching it between the tips of my fingers, scatter it like this. The seeds fly a couple of yards, so an hour is all it takes to sow a quarter-acre. My clover seedlings right now are about an inch high.

A week or two before harvesting the rice, I take anywhere from about four to ten quarts of barley seed, place it in a basket, and scatter it over the field. This also takes about an hour. After the rice is harvested and threshed, I scatter the straw back over the field. The straw should always be returned to the field uncut. The more disorderly the scattered straw the better. When I told one university professor to simply scatter the straw, he cut it and lay it neatly over the field. Naturally, the barley didn't germinate. When he asked me why not, I told him that he had laid the straw down too tidily. You don't arrange the rice straw carefully, tie it up in bundles, or cut it up. It has to be scattered loosely. The barley grows up through the spaces left between the straw. The layer of straw should be about so thick. It's okay to scatter all the straw back on the field. When I tell researchers that, given the size of the test plot, one can scatter 200 pounds of straw here, they are shocked by the very idea and say flat out that that's too much. But it is best to return all the straw to the field. It doesn't matter whether you've got three, four, or five hundred pounds of it. Just go ahead and scatter it all . . . right over the barley shoots.

And here's something else for those who want to try a different approach. Take your rice seed out to the field in a basket and scatter it sometime from mid-November to January. First you've got clover, then barley growing. Then you sow rice seed on top of this and stay out of the fields until it comes time to cut the barley. One hour to sow the seed, two or three hours to scatter the straw, and barley seeding is over. You should keep out of the field until harvest time. However, if your aim is to reap more than 20 bushels per quarter-acre, then you might go in once and try applying four to five hundred pounds of chicken manure. Of course, with all the chemicals used to raise poultry, perhaps it's better not to use chicken manure. Human waste these days is also polluted, but the fields do get fertilized by the sparrows, moles, and other small creatures that visit or inhabit the fields. Human wastes should definitely not be overlooked as a valuable resource. If you've got to have yields of over 20 bushels, then go ahead and use chicken manure until a better alternative comes up. It's just a matter of time anyway. But don't apply anything else. Take home the grain, but make sure to return all the hulls to the field. And do not take out so much as a single barley or rice straw.

When the prefectural testing stations and universities lay their hands on it, this simple process of scattering a single straw can become quite complicated. Fifteen years ago I said that the straw should be scattered uncut. But this was laughed off by the folks at the Ehime Prefecture facility as just too crude. What they did was to chop up the straw into small pieces with a cutter, carry it out to the field in baskets and scatter it. This takes a whole day. It's entirely unnecessary, but if you feel you must do it this way, then go ahead and try. Of course, it will take you three years instead of one to realize that there's no need to chop up the straw. In the third year, the researchers at the Ehime station told the farmers, "Sure, that's what Fukuoka said, but it looks as if cutting the straw into three pieces is just about right." So they did this for three years and began realizing that tossing the straw back onto the field in even longer pieces would have been better. All told, it took them nine years to come around to my way of thinking.

They say that those who study plant pests and diseases should know better than to be scattering straw. That was my specialty, so I remember well how before the war farmers in Hokkaido were all told to burn their straw because of the pathogens remaining on it. Everyone thought that straw was crawling with disease. The way the authorities saw it, scattering the rice straw back onto the fields was tantamount to sowing the seeds of blast and sclerotium disease, so they instructed the farmers to either burn it or bring it home and turn it into compost. Incentive money was even handed out to farmers for composting. Every month a competition was held. The judges included people from the local agricultural testing station. A hundred pounds of straw brought in from the fields and converted to compost weighs two hundred pounds. That straw had to be taken home and turned six times to prepare the compost. Try composting straw like the farmers in Hokkaido and you'll see what it's like to turn hot, steaming compost with a pitchfork in a tiny shed. Young people today are totally incapable of doing this. It was hellish work, yet the farmers went at it earnestly.

It took me ten to twenty years just to say, "All you have to do is scatter the fresh straw back on the field." But that's not enough to convince the experts. "Don't you have pathogens on your straw?" they ask. Take a look for yourself and see whether I do or not. Look, I was a plant pathologist myself so naturally I tried isolating the agents of rice blast disease and other infectious pathogens from fallen leaves. But I never succeeded. The fact that I, who was good at isolating pathogens, failed to do so here could only mean that the pathogens were dead. If they were dead, didn't this mean that it was okay to return the straw to the fields? So that's exactly what I did. But it took five years for other plant pathologists to come to the same conclusion. And it took another five years for the soil and fertilizer experts to discover that it's not necessary to spread manure when scattering the field with straw. So five years, ten years passed and eventually the folks over in the pathology division of the research center got together with those in the fertilizer and the plant husbandry divisions, ran some more experiments, and compared results. And, lo and behold, they found that rice straw was fine by itself. It took seven years for them to finally catch on to the fact that rice doesn't have to be transplanted. All this time I've been waiting. It doesn't have to be only me running these tests. I'd be delighted if others were to try some of these things out for themselves. But unfortunately, government research centers and universities don't do studies of this sort. They work only in the opposite direction.

There was once a time when one individual considered everything—pathology, the plants, the soil, and fertilizers. But today research has become fragmented into narrow disciplines. People say that if you combine all the data, you'll get the total picture. But you can't first take something apart, then fit all the pieces back together again and expect to get what you started out with. Suppose that, in order to know a flower, you take it apart, plucking off the petals and stripping the leaves from the stem, then have pathologists and botanists examine it. Once you are through, if you then try to gather all the parts together and reconstruct the original flower, you'll find that it can't be done. But that is exactly what researchers are trying to do. And that is why they are drifting away from the true essence of things and from nature.

Everything would be all right if what the agricultural scientists and technicians are doing were in the best interests of the farmer, but take a look at the universities we have today. All you can see are buildings. There are plenty of lecture halls, but where do you see any farmers? Farmers have nowhere they can go for consultation. In the past, farmers were able at least to go on over to the university fields and ask how something should be grown. But nowadays, you take some rice leaves with you and wander about without any idea of where to go.

There is no such thing as a non-polluting pesticide. Farmers today are spraying so-called low-toxicity agents. "Low toxicity" does not mean only slightly toxic. What it really means is toxicity that is not immediately apparent to the human eye, toxicity that is difficult to discern through chemical analysis, even when examined at the microbial level.

The low toxicity of these pesticides provides a degree of latitude in their use. But, if anything, that just makes them more harmful because new types are developed one after another in rapid succession, complicating matters. It is the most talented chemists who develop pesticides such as these that ordinary people are unable to avoid as manifestly polluting. Science and technology strike people with one hand and help them with the other. Both tendencies can be found in the universities. Each college claims to have its own policy and objectives. Things have clearly gotten out of control, but who is going to bring them back under control? And in what direction? To tell the truth, I think that we have entered an age of confrontation between science, philosophy, and religion.

Nature as Teacher

One of the farming methods being studied lately in Japan is the same organic farming that has gained such a wide following in Western countries. This is Eastern farming on the same plane as Western farming. Agricultural testing stations and universities have done quite a bit of work with organic methods.

With the soil dead, people are saying, "Let's add organic material." But if this were all there was to it, then it would suffice to return to the primitive farming methods of the past. On the surface, organic farming and natural farming appear similar, but consider this carefully before you decide one way or another. It took me twenty to thirty years to be able to say, "Scatter straw onto the fields." Scattering that straw makes composting unnecessary. A thousand years ago, peasants in Japan did not till their fields. In the Tokugawa Period (1600–1868), peasants began shallow tillage. Then in the late nineteenth and early twentieth centuries, Western methods were introduced and deep plowing was adopted throughout the country. I am in the process of returning to a method of farming that does not require tillage. One could say perhaps that I've returned to the past, but natural farming is not just a simple reversion to primitive methods.

I don't plow my fields, but I do sow clover. This is the easiest way there is to grow rice. With the arrival of spring, the clover grows thick and fast. I sow rice seed in this clover and later flood my field to weaken the clover and favor the rice. I then drain the water and leave the field to itself. The rice, barley, and clover work the soil biologically. I don't plow, but lay straw over the surface of the field. The straw and clover do more for the fertility of the soil than large tractors. So this is certainly not a primitive farming method from the past. It may seem primitive if all you pay attention to is the word "no-till," but it is in fact a biological method of farming that uses plants and animals rather than heavy machinery. If you think of this as a means for raising soil fertility using microbes, as cultivation with plant roots, then it becomes the most advanced science.

Natural farming is a method that actually goes a step beyond science. The proof is that over these past twenty years I have not read a single book on the topic, and yet have managed to stay at the forefront of rice cultivation practices in Japan. Try comparing the findings that I reported ten years ago with what agricultural testing centers have been doing thereafter and you'll see that most prefectural centers are now doing what one lone farmer tried a decade ago. And they're arriving at the same conclusions—but, of course, ten years after me.

My purpose here is not to boast. I would simply like you to know that I walk always the shortest road. By this I mean that my teacher is nature itself. Nature is always perfect in every case. I have had a good teacher. Other people have broken nature down and looked at it only as small fragments, so what they learn is incomplete.

224

Because they have a bad teacher, despite their diligent studying and application, they reap little from their efforts. Science never does any more than mimic a virtual image of nature that exists only in the human mind, so what it grasps is only an incomplete and inferior imitation of the real thing. I can assert here without the least doubt that anything created by man with scientific knowledge will always be inferior to nature. When one realizes just how wondrous a thing nature is, one can only bow to it in humble acknowledgment. The moment that we become humble before nature and renounce the self, the self shall become assimilated into nature and nature shall allow it to live. Even a small ego becomes capable of summoning great strength. It is enough merely to know this road and walk it each day.

*   *   *

I delivered this lecture before the Sekai Kyūsei Kyō more than ten years ago. Today things are a little different on my farm. About four years ago, I stopped having students stay at my orchard huts and retired into a solitary life of seclusion and study. My son and his wife care for half of my orchard, while I manage the other half and a little over an acre of rice paddy, where I do as I please.

Depending on how you look at it, this orchard may seem disorderly wilderness now, but I think that in the space of a few short years it will gradually take on the form of an integrated farm, something which I await with great anticipation.

As for the rice field, for thirty years now I have employed a direct-seeding, no-till, rice/barley succession in a green manure cover. But because it has become clear to me what methods can be practiced anywhere and by anyone, over the past few years I have made some significant changes in my methods.

What Is Nature?

Nature Is Unknowable

If someone were to ask me what nature is, I think, quite
honestly, that I would be unable to give a satisfactory answer.
And yet, everyone makes free use of this word *nature* without
giving much thought to it. I too talk of the form and spirit of
nature as if I knew it all, but deep down I feel that the mere
mention of such a notion and the very attempt to express and
describe nature are the root of error.

Perhaps a way of putting this that approaches more closely
to nature would be to say that there is no form or spirit to
nature. When one asks what nature is, this conjures up images
of fields and mountains rich with vegetation, of the cosmos
and heavenly bodies, of the vast reaches of the wild. Men of
religion perhaps imagine the sort of universe mentioned in
the Book of Genesis as the essence of nature.

In general, people think of nature as the natural phenomena
perceived by the natural sciences; they believe that the plants
observed botanically are nature itself. In literature, we use the
expression "to each according to its own," and treat the state
of natural flux as nature itself.

But I believe that there is no way of expressing the true
essence or state of nature. No matter what words we choose
to use, we are only able to discuss botanical nature, to
describe the form of nature from a cosmic view. Being unable
to come into direct contact with the essence of nature, we end
up circling round and round its periphery, conjuring up
associations.

Even the image of nature that people form in their minds,
of a world left entirely to itself, unaltered and untouched, is

itself a vision anchored within the unnatural human intellect and human action. Rather, the tabula rasa state of nature that transcends nature as seen and considered by man and which arises by abandoning all notions of nature is closer to true nature.

However, because "tabula rasa" is merely the opposite of what is not tabula rasa, expressing things in this way does little good. No way exists to describe nature as it truly is. The best I could do would be to say that if one casts off everything, absolutely everything, from human thought, what emerges thereafter in one's soul—that indefinable something that one apprehends after having transcended even the light of which the *haiku* poet Basho wrote in his poem: "Oh, how splendid! The sunlight on the young, green leaves"—that could be called nature. I believe that no better explanation is possible.

It should be clear then that explaining the word "nature" is impossible. Yet, in spite of this, people seem to think that by studying nature in all possible lights, by combining the nature known to the naturalist, the nature known to the man of religion, the nature as seen by the artist, and the nature as conceived by the philosopher, that one will eventually come to understand nature. But true nature cannot be broken down and analyzed, then reassembled and understood as a collective whole. Understanding does not come through analysis and reflection. People believe that the form of nature can be broken down from different perspectives and grasped through synthetic judgment, but this just is not so. God too shows us many different faces, yet the moment that we distinguish between these and serve up commentaries on each, God ceases to be God. In the same way, nature becomes unnatural as soon as it is depicted in a painting. It is conceptually classified, torn apart, and kept at a distance, becoming something altogether different. And of course, once nature becomes estranged from God, it is no longer able to return to its original self. This is the same as the estrangement of man

from God (or the banishment of man by God). The stronger
the conscious desire to know, the further man becomes distanced from nature.

The nature that transcends the discriminating and relative
thought of man is true nature, but that which lies beyond the
relative world of man cannot be expressed with human
language.

Consciously or unconsciously, man cannot comprehend
nature. Even if one were to venture that the nature seen from
a point of non-discrimination lying beyond the unconscious is
true nature and God, such a point cannot be reached by man.
Unfortunate as it may be, this means that man does not know
transcendent nature and God. Thus, although we may try
explaining nature, all we can do is to explain that nature is
something that cannot be expounded upon.

Crossing Mountains and Valleys Gets You Nowhere ——

Thinking that he must climb a mountain to know it, the
mountain climber climbs the mountain. But in fact, to know
the true mountain, one must see it from a point of remove
that transcends the mountain. Man can scale a mountain and
stand on its summit, but he cannot climb above and beyond
it. Unable to grasp the mountain in its full aspect, he is
content with having seen only one small portion and climbs
back down.

Mountain climbing, going to the beach, listening to birds
singing—all these seem to be ways of capturing a bit of
nature. But no matter how many such recreations one gathers
together, they do not add up to a true understanding of nature. All this does is to blow up the conceptual notions of
nature as interpreted by the natural sciences. The accumulation of discriminating knowledge only deepens confusion.
It draws man away from nature so that he no longer understands it.

There is no way of expressing that mountain which goes beyond a mountain. Nature can only be understood with a nondiscriminating heart. To see a mountain one must go beyond the mountain; to see the sky one must go beyond the sky. One must see the mountain from the world of the sky that emerges only from a philosophical perspective. This is not the view that one sees from the summit after having climbed a mountain. The sea of clouds, the valleys and lowlands far below, and the distant mountain ranges that one sees after climbing a high peak are all a part of the temporal world. Viewed from the celestial world, even the tops of the clouds lie beneath the sky. The sky above the sky cannot, after all, be transcended. We constantly turn back to the conscious world, from which we cannot pass to the unconscious world. That is why ours is an empty world in which we know nothing.

Even abandonment is far from being natural. When man alters and destroys nature, then throws it aside, that is abandonment. Being that he knows not real nature, man has no way of telling whether nature abandoned and left to itself will return to its original form or whether it will hurtle in the opposite direction toward anti-nature.

Perhaps one could say that the only way of approaching nature is to notice that man understands nothing—neither nature nor anti-nature.

The Innocence of Children

I often say that it is best to think as little as possible. People find fault with this. "If you don't know clover from sorrel," they object, "then you can't practice natural farming. You've got to know when to flood and drain your rice field, when to scatter straw, and whether or not to sow radish seeds in their pods."

I prefer to think of this as spontaneous knowledge that

existed before scientific knowledge and serves only as a crude
imitation or explanation of what nature does naturally. One
does not need to learn knowledge from science; it is enough
to learn (that is, imitate) nature. When I sow radish seed
mixed with clover and try explaining that green-manure
cultivation allows the radish to be grown without fertilizer,
this puts me in a ticklish spot because others then think it
necessary to know that the clover is useful as a green manure
plant while the look-alike sorrel is not, and that there are
many varieties of clover, some of which serve the purpose at
hand better than others. But that is exactly the same as
thinking that to know God or the Buddha, one must know
their many faces; that unless one knows the sutras well, one
cannot attain Buddhahood when one dies. Yet, just as attaining Buddhahood is possible without knowing even a single
book of sutras, it is not true that one absolutely must have a
knowledge of plants in order to practice natural farming.
Sutras are used to explain what the Buddhist deities are. Like-
wise, since most people know something about plants, the
use of botanical knowledge serves as suitable material for explaining natural farming. Knowledge appears to be convenient
for expounding on Buddha and God, but this serves more as
a hindrance than a help to one who would become a true
man, an enlightened man.

For example, it is not necessary to tell a child, "This here
is wood sorrel. It looks like clover, but it's not." A child does
not understand and has no need for such botanical knowledge. Teach a child that clover is a green manure plant and
that pearlwort is a medicinal herb useful for treating diabetes
and the child will lose sight of the true reason for that plant's
existence. All plants grow and exist for a reason. When we
tie a child down with petty, microcosmic scientific knowledge,
he loses the freedom to acquire with his own hands macrocosmic wisdom. If children are allowed to play freely in a
world that transcends science, they will develop natural
methods of farming by themselves. It would have been better

not to have known enough to distinguish between pearlwort and clover.

Natural farming is always incomplete. Even my natural way of farming is still only the very beginning. All I am saying is that the wood sorrel may sometimes get in the way on a small farm, so replace it with clover. But when a grander, more "macro" form of natural farming is established, anything at all—wood sorrel included—will be allowed. This has to be brought to a point where nothing in the world does not serve its purpose.

Just look at children. A while back, a group of forty or fifty kindergarteners visited my farm, where they ran about, playing and filling the orchard with their laughter. They found seeds on the large, thick-stemmed burdock growing there and had fun throwing these onto the clothes of other children. Here then, just as the birds and wind carry off maple seed, the playing of these children was helping to sow the burdock seed. They had become a part of that natural vegetable garden.

If you try throwing weak, spiritless youths into the hills to fend for themselves, they'll manage quite well. This means that they know how to live. An adult cannot practice natural farming without distinguishing between clover and weeds, but young people grow and search for food in the midst of nature as they find it. So I do think that scientific knowledge is necessary, but only while natural farming remains immature and confined within the realm of scientific agriculture.

In a sense, my book, *The Natural Way of Farming*, is an attempt to explain in "macro" scientific terms that one can practice natural farming without "micro" farming knowledge. For instance, specialists talk of the need for fertilizers, pesticides, and plows. To them, I reply, "Look, if you do things this way, maybe you won't need any of those." The argument that crops can be grown without fertilizers, pesticides, or tilling is not without grounds. If I were to explain it scientifically, here's how it would go: Although I don't plow mechanically, plant roots and small animals work the soil biologically. In fact, this biological "tilling" works the soil deeply. The practical limit for tillers in Japan is about six inches. Even the large machinery used in America does not plow deeper than a foot or so. Barley roots and moles go deeper than this. When left alone, the soil deepens and enriches naturally. Tilling the soil mechanically only kills it. Instead of doing what he set out to do, by plowing his field, the farmer is just ruining it.

When I explain things this way, people say, "All right. I understand that the natural farming technique of no-till direct-seeding is the shortest path to true enrichment of the soil." Yet even with my explanation they fail to see that science serves no useful purpose. They believe that I've developed an advanced form of natural farming by applying scientific knowledge to the primitive agricultural methods practiced by farmers of the past.

No, natural farming is neither a method that returns to the ignorant past, nor a method developed on the basis of scientific knowledge. As I have said countless times, natural farming was born suddenly one moment almost fifty years ago. I began with the conclusion that tilling, fertilizers, and weeding are not necessary. This was not a conclusion drawn on the basis of knowledge in the agricultural sciences. It arose from the religious view that all is unnecessary. The starting point was the philosophy that rejects the human intellect and denies that objects and human actions have any value. Thus, although natural farming remains of course incomplete, it is gradually moving, and must move, toward the rejection of science. Natural farming arose from the same ignorance as that of children. It may well be that only ignorant children will develop new natural ways of farming.

When I talk of the non-value of all things, this applies as well to the thing called nature that is observed by man. That too is without value because it has no bearing on the happiness of man.

The first time I met Professor Keiichi Sakamoto of Kyōto University, I told him: "Professor, why don't you set up the study of *Mu** economics? I'm a farmer, so all I've been able to do is demonstrate the theory that all is worthless from the standpoint of farming. Since your specialty is the principles of economics, you ought to set up a field of economic theory which throws out current economic notions that things have value, and is founded instead on the principle that things *do not* have value. It should be possible to establish a new science of economics that totally rejects Marxian economics, the *Das Kapital*, and modern economics."

I said this ten, maybe twenty years ago. At first, he seemed to have his doubts, but now he appears to be advancing slowly in that direction. Of course, I imagine that it will be very difficult to go so far as to totally reject the value of things.

The first questions to arise when trying to establish a system of *Mu* economics will probably be whether things have value and whether the conditions and premises that make things necessary are truly essential to man. The reason we think that a glass of water has value is because conditions exist that give it value. In the desert, this water has value. But under a different set of circumstances—in a flood, say— water has no value at all. Depending on whether it is an hour before or after a rainfall, the water may have value or the value that it had may suddenly vanish; the value may rise or fall. So what then is the value of things? Is there intrinsic value in this thing called true nature, or does man uncover value in the nature within things? Or is value perhaps a product of the relationship between both of these? What determines the true value of water? I believe that the economists should hurry and set up an economy of happiness that starts out from this point.

The ancient Greek philosophers said that it was necessary

---

* Nothingness. Intended here in the sense of "non-activity" and "non-value."

only to know the five elements—earth, water, fire, wind, and light—but I suspect that they really wanted to reject even the five elements. Far less, then, did the need exist to study and acquire four hundred elementary particles and elements. Mankind today sees value in science, but just what is this value? I sense that the outcome of all that research is that we are toying with four or five hundred elements in a world that has not the slightest connection with true human happiness.

Japan's Hideki Yukawa, for example, won the Nobel Prize for his research on elementary particles, but I have my doubts as to how much good this has done in bringing happiness to man. Yukawa went deeper and deeper into the tiny world of elementary particles and mesons, where he discovered a vast, microcosmic universe. People may have thought that by gazing intently at this extreme, he would come to know the macrocosmic universe. But realizing that there is no "large" in a macrocosm and no "small" in a microcosm he was unable to proceed any further and sought instead the salvation of the world of Buddhism. Yukawa came to see that we cannot tell large from small with science.

The fact that there is no intrinsic value to objects in nature and no need for research seems to have a direct bearing on the happiness and unhappiness of man. Let me say simply that true happiness does not arise from things. I may be understood here as saying that because the nature seen by man is not true nature, false natural bodies cannot have true value.

Washing the Naked Body

Although I have on occasion told people to practice Zen meditation, I often urge the foreigners who visit my farm to visit the Dōgo Hot Springs nearby. "Since you've come all this way," I tell them, "you ought to stop by the hot springs.

Sightseeing is okay, but I suggest you get into the baths at Dōgo and take it easy for a day. Look, you're wasting two months anyway, so before you head back, stop in at the baths and cleanse your soul."

Dōgo is Japan's oldest hot-springs spa. Created, according to legend, by Ōkuninushi no Mikoto, the Shinto deity, it has been visited by many distinguished people, from Prince Shōtoku to Emperor Hirohito. Poets such as Issa, Bashō, and Shiki have come here. Modern writers who have stopped by include Natsume Soseki. Priests such as Kūkai, Ippen, Kūya, Bankei, and Mokujiki Gogyō have come here, as have all the major Japanese thinkers. So going to the spa, reading the biographies of these thinkers on the second floor above the baths, and trying to imagine what was running through their minds as they were enjoying the waters can make for an enjoyable time.

What I like most of all at Dōgo, however, is just stretching out on the granite floor of the baths. In a stone bath, your back warms up and makes you feel good. The sensation when you lie down with your arms and legs stretched out is indescribable. The whole body unwinds. As the muscles relax, you become comfortable and free. The heart too loosens up and relaxes. You become free and uninhibited. This loosening of the body is, I believe, the road to oneness with the Buddha. It is a shortcut to the Buddha—which is why I'm always going over to Dōgo with the excuse that, "Zazen is fine too, but one can also attain perfect serenity by stretching out at a spa."

I have no objection to those who earnestly practice zazen on a tatami* mat, but the outcome is the same as the farmer who cuts his weeds. I wonder if one can be at rest by joining one's hands, crossing one's legs, and staying perfectly still. I would think that in moving one's arms and legs, one's attention would be diverted to this, emptying the mind and achieving the same result as if one were to sit quietly. This too, then,

---

* Rice straw mat used as flooring in Japan.

could be considered one way of practicing Zen. It is easy enough to say, "Sit quietly in the lotus posture, face the wall, and empty your mind of thoughts." But that is a tall order. The more you tell people not to think, the harder they think. That is why, there being no alternative, trainees are often told, "Count to yourself—one, two, three...." But in doing so, one becomes fettered to the very act of counting. Counting is tiring and a waste of energy, so one doesn't continue. Using one's head to count is not permitted, but neither is it acceptable to let the mind wander and forget to count. No matter how hard one tries, one fails. Indeed, the greater the effort, the more likely the failure.

Someone bound by his own awareness and intention to practice *zazen* is unable even to move. One would probably feel freer by forking out the pittance charged and entering the hot bath to stretch out. After all, this can be done at one's ease. That is what I suggest: easy, comfortable Zen meditation. I myself don't say that *zazen* is bad, nor do I strongly urge anyone to do it. To begin with, I've never done it myself. All I have done is to go to the spa at Dōgo, lie down in a hot bath and feel . . . good. There is no more to it than that.

But that probably is as far as people should seek to go. People should live easily and enjoyably, and die a peaceful death. For this, a natural body is best.

Meditation has become quite popular lately, but where it involves mental concentration, I consider this a form of brainwashing that is, if anything, a dangerous road. Zen, I believe is oriented toward escape from the world of ideas. It is foolish to be tied down to things such as the soul and malevolent apparitions that don't exist at all.

The Natural Body

One person that made an impression on me at a meeting I attended was Dr. Keizo Hashimoto, a chiropractor from Sendai. What he does when a patient comes in is very simple.

He has the patient lie down on a cot and just tugs on his legs and arms a bit, that's all. What he's after, he said, is simply to make the patient comfortable. "If the right arm hurts when the patient raises it, then I have him shake the left arm in the other direction. I just try to make him feel good. If he can get rid of the pain, that's fine." In other words, it is enough to relax. Dr. Hashimoto says that "people ought to live pleasantly and die pleasantly." Those are wise words, indeed.

In the same vein, a physical education professor at Nihon University said that all one needs to do during exercise is to make the body perfectly limp and let the limbs dangle. I touched his body and was astonished to find that although he was just standing there doing nothing at all, inside, his body was rippling. His inner organs were exercising by themselves. "The human body acts as if it were made of Jello," he explained. "It's like a leather bag filled with water. If you shake it, it even sloshes about inside. Just let the arms and legs and everything else dangle. No one has to put himself through a hard workout." He told the audience to build bodies as flexible and strong as a whip.

The way I see it, techniques in Oriental medicine such as *shiatsu* (finger-pressure therapy), acupuncture, and moxibustion are gradually being unified and simplified. In spite of what everyone says, I get the feeling that things are moving in the direction of doing nothing. I have long maintained that sports today which insist on rigorous training to toughen the body are a mistake.

Malady always arises from what is unnatural. Mental disharmony with nature hardens the mind, while unnatural physical care stiffens the body. By continuing to endure such conditions and failing to exercise moderation, one eventually meets up with serious illness. To recover, one must lighten the heart and live easily—without strain. That, I think, is what it is all about. If you do nothing and simply return to the natural body of an infant, then your health will improve. If you have become neurotic from thinking too much, then by reverting to the state of an infant who thinks of nothing, the neurosis will be cured. However, when one is told to become an infant, the mind does not return, and when one is enjoined to become a fool, one is not bright enough to do so. Should this happen, the only choice is to allow that person to amuse himself with total abandon or to cast him into the mountains and leave him there to himself. In this way, willing or not, he becomes a fool. He will resign himself.

For the first day or two, the folks who come to stay in my orchard put up with things, but after a week has passed they begin to lose track of the time and come to me asking what day of the week it is. What was to have been a week turns into two weeks, three weeks. After two weeks, they feel as if they could stay on for two or three months. And when three months have passed, staying for a year or two becomes quite easy.

This certainly is one fashion in which therapy could be applied for illness.

Natural Farming Today

Some Are Returning to Nature ────────────

All aspects of modern life—learning, politics, economics—are moving away from God. For many years now I have tried to hold up natural farming as the shortest road to God and as an example of a way of living to which people should return. But as far as the diffusion of natural farming is concerned, things look quite hopeless. People have been getting excited over natural foods and organic farming, and the "back to nature" movement seems to have really caught on in Japan. But compared to thirty-five years ago, it appears to me as if people are simply moving farther away from nature, as if our society and civilization is actually crumbling. The reason things seem so hopeless lately is that although the Japanese say they are "returning to nature," they are not.

In contrast, when Westerners talk of returning to nature, they tend to go right ahead and do so. A short while back, a young Briton who came to stay at the huts in my abandoned orchard told me, "I was unable to find peace of mind anywhere in Japan. But after arriving here, I've been able at last to breathe freely." He asked no questions about the techniques of natural farming and did nothing while he was here. Even so, foreigners return home content. And when they go back, they put natural farming into practice. Many will read a single book—*The One-Straw Revolution*—and immediately set out to try my methods. During my travels abroad over the past several years, I met quite a few individuals who had done precisely this. Why is it that the Japanese are unable to do what Westerners can? The Japanese have for so long practiced an intricate and methodical type of farming with pocket-sized plots; they seem to know nature best of all and would appear to have the least trouble returning back to it. Why is it then that the Japanese today, despite their talk of leaving the cities for the country, are unable to settle down and return to nature? They go no further than mere imitation, making do for the moment but failing to continue for more than three or four years and eventually drifting out of farming. But Westerners get right in there and stick with it.

When I ask visitors from Britain what they think of Thatcher, they generally say something like: "No, thank you." Then they'll add, "No matter how hard one tries to get something going at home, English culture and civilization is doomed. There's no hope for recovery there." So they come to visit my farm, look around a bit, and feeling convinced that this works, strike off for Australia to farm the land. A goodly number head for Australia or New Zealand via my farm. What they are saying is that Western Europe is beyond hope anymore.

The Japanese who come from Tokyo and the other big cities saying that they despise urban culture and want to get out and return to nature share the same desires as these foreigners. Their motives are pretty much the same. Why is it then that only Westerners are able to get right in there and act out their sentiments? I think one explanation is that Westerners know full well the errors of Western philosophy. Although the Japanese are familiar with Eastern philosophy and Buddhism, these have become merely a body of concepts to them, they have forgotten the soul of those teachings. Despairing at Western philosophy and beset with doubts over Western culture, Westerners decide to get out of this mess and turn to Eastern thought and Buddhism. Once they have so resolved, they plunge right in. The Japanese, on the other hand, remain immersed in Japanese culture and the lovely nature of the past—which is why, I suspect, they are unable to extricate themselves.

I think another reason is that people abroad who look at

my fields or read my books are more often in a position to try these methods out themselves. It is more difficult in Japan to act upon such a resolution because the society in general is less tolerant. One is subject to criticism and obstruction from all sides: "You can't do that because it runs against the recommendations of the Agricultural Cooperative Associations" or "Government policy makes what you propose to do impossible." But I've heard that in Australia, land prices are only about a hundredth of what they are here in Japan. In fact, I've had young visitors to my farm who, nursing aspirations of becoming natural people, work and save up money for a year, then buy some land in Australia and become farmers.

Land is cheap so one has no need to worry if yields drop a bit. But in Japan, even a small drop in yields draws laughter from the surrounding farmers; and if you don't spray your fields you are upbraided for being a nuisance to the farmers around you. So one pays all one's attention to things like that and ends up being strapped down by the residual feudalism of agricultural communities. Even so, the Japanese today are lacking in determination. They are too skeptical and have no strength of conviction.

I think that the fact that foreigners are unable to speak the Japanese language when they come to my farm is a plus. When young Japanese people decide to visit my farm, they come with the desire to learn about natural farming. This is apparent in their attitude. They come, bringing what they know, with the intention of hearing what someone else has to say and in this way adding to their store of knowledge before starting to farm naturally. Westerners, however, come after having read only *The One-Straw Revolution*. In that book, I wrote that the intellect serves no purpose, that it should be renounced. Those who come after having read this are prepared to do just that. They empty their heads and, in the fashion of Buddhist devotees, abandon themselves entirely to natural farming. That is why they don't make an effort to learn or indulge in mental gyrations. But the Japanese do. They come and say, "Mr. Fukuoka, teach me something." It is not so bad when they look at me with their sharp, watchful eyes and challenge me with, "What does naturally grown radish taste like?" and "Is a radish Buddha or nature?" Some think that all that is required to practice natural farming is the answers to such questions as: "Where can I buy clover seed?" and "How much seed will I need?" I think this is the main difference between the Japanese and Westerners in their attitude to natural farming. Some Westerners who read my books and believe in what I say come all the way from Europe, America, Canada, Mexico, Australia, or wherever. They tell me that they have come because they believe that what I wrote is "correct." By correct, they mean that they think this is the truth. When they read that one must "abandon knowledge and the intellect," they firmly accept this as the truth without vacillation. That is why they are able, right from the start, to become tabulae rasae. Drinking the clear spring water and experiencing life by the hearth in my orchard huts, they treasure this and leave, saying, "That was great. I'm really glad I came." There is a purity of spirit there.

So the way in which Westerners and Japanese abandon knowledge and the status quo differs. The Japanese immediately turn around and ask, "What is natural farming?"

One spring day a couple of years ago, a young man from a religious organization called Tendai Kyōdan arrived at my farm. "I felt that something was wrong," he explained. "I was put on a pedestal as leader of the sect, but I was nothing more than a figurehead. So I withdrew and came here. Mr. Fukuoka, I'd like to help you out on the farm and work to spread natural farming around the country."

"If you're saying 'I've renounced everything,' then do so," I answered. "Isn't it enough to be merely a farmer?" I didn't like what he had said about helping me out and spreading natural farming. "You keep saying, 'I've abandoned everything,' but you haven't done so like a Westerner, have you?

If you've shed your past, then shed it all. Here you've come, trailing behind you ten, twenty followers. Why? And you talk of helping me out with natural farming, but are you capable of working as a farm hand? Isn't it a fact that you don't know the first thing about the hardships of a farmer? Have you ever done any farming yourself?"

"No, I haven't."

"Well then, why in the world do you think you can help me and spread natural farming? I've been practicing this for forty, fifty years, but I'm still not prepared to teach it. And I don't even have a single follower. Instead of pleasing me, your offer irritates me."

Being the leader of so many believers, he was quick to understand. When he left, he said, "I'm returning to Hokkaido, where I'm going to start over as just a farmer."

What helped convert him was the following passage from my book *Kami no Kakumei** (The God Revolution):

> Compelled to speak, God says nothing.
> Unable to say anything, man speaks.
> Man is ignorant of the truth, but expounds on all fields
> of learning.
> God is versed in all things, but expounds on nothing.
> God takes no action, yet creates all things.
> Man does all things, yet creates nothing.

Why Doesn't Natural Farming Catch On in Japan? ——

I am sometimes asked why it is that no other farmers nearby have thought to try farming the way I do. A farmer does the same thing year after year, so even a period of forty or fifty years passes by in a flash. I don't have the feeling that I've been doing this for a long time, and I think that those in my neighborhood probably feel the same way. The fact is that we

---

* An English version of which will be published in the future.

have not had the time to indulge in idle conversation on the walkways between our rice paddies. I've never given a talk in town. The farmer next door simply says, "Muggy today, isn't it?" or "Chilly weather we're having," as he walks by on the way to his fields. And when he does take a sidelong glance at my field, he notices that it's always full of weeds, that it's bumpy and uneven with high and low spots, that sometimes it's flooded and sometimes not. To him it looks like a mess— no order at all. There is no denying either that I have often had bad or irregular harvests in my fields. My neighbors, feeling sorry for me, just pass by, trying not to look at such a sorry sight. Each year I have tried something new, experimenting with new methods of growing rice. So, to the farmer who treasures stability, I am a heretic of sorts—a troubled soul. No one dares change his way of farming and follow my example. More precisely, no one has even the slightest idea of what I've been doing and how to do it. That is because I am always changing my methods; I never do exactly the same thing twice. To tell the truth, I have yet to provide results good enough to inspire the people of my community.

I imagine that the local people have heard something about natural farming from the newspaper and television coverage it has gotton. But they have their pride. They think, "There isn't any way someone can grow crops without plowing, fertilization, and pesticides. Look, we know. We've been growing crops now for decades, for centuries." So when they see photos of *daikon* radishes growing beneath unsprayed mandarin trees, they don't think these grew by themselves. And if these had, they would discount it as a chance occurrence. Not a single villager has visited my hilltop orchard and seen for himself the radishes growing there. Those in my immediate neighborhood have seen, and yet not seen, my fields. It is like a bird singing that no one hears. They are always seeing my field so they feel they understand; that is why they don't look at it or speak about it. The fact is that no one is moved even to ask about it.

The little natural farming of rice and barley or wheat being

practiced today is done by people a long ways off. No one
from my village has ever bothered to visit my field or orchard. Those who come generally do so from faraway places
such as Okinawa, Hokkaido, and abroad. That is because,
to the farmer's eye, the faults are all too obvious. Farmers
zealously pursue farming methods known to be reliable and
complete. But if a method does not produce well-ranged and
uniform results that are pleasing to the eye, if a field is
irregular and has weeds growing among the grain, they
normally pay it no attention. In a sense, the people of my
village know better than anyone else the rigors of natural
farming.

Although I may think I am using the same method every
year, the circumstances always differ, resulting in variable
harvests. While I had good yields in some years, in others,
yields were bad. In general, if a farmer sees one spot that is
not doing well in my fields, he figures that natural farming is
unstable and doesn't even give it a try. A farmer that has a
bad crop one year suffers for two or three years after that.
Or worst yet, if he stumbles so much as once, a farmer these
days may become unable to pay off his loans. This fear of
crop failure has become ingrained after three or four hundred
years, so farmers dislike nothing more than an irregular harvest. No one is as methodical and unbending as the farmer.
As for me, I have done just the reverse by trying every possible type of irregular cropping under the sun. So it is no real
surprise if local farmers and I have gone our separate ways.

Agricultural specialists all know that it takes a minimum
of ten years to bring about a reform in even a minor farming
technique. Although you might suggest to a farmer that he
try a new method out, normally the local agricultural research station has to try it out first. Then the prefecture
recommends it and the technical staff who disseminate farming improvements make the rounds of local farms to explain
the technique to individual farmers. All this takes at least
ten years.

No wonder then that it took ten to twenty years for the method of seeding barley on level rows that I proposed to catch on around the country. No one paid this any attention as long as I called it "pared seeding," but when the folks over at the testing stations ran tests on it and mechanized it, renaming it "full-stratum seeding," this caught on at once. It is not that no one knew anything about the method or had never heard of it. Farmers were aware of it, but they were waiting to see what the specialists over at the research stations said and what sort of advice the prefectural people would give. No one watches his neighbor so carefully yet guardedly as the farmer; no one does as much research as the farmer while appearing not to. Most farmers think that when the experimental stations, the agricultural cooperatives, and the farming guidebooks start saying something, then it is safe to go ahead. They believe that going so much as one step ahead of the others is a prescription for failure and will leave one out in the cold.

Farmers in Japan today receive their instructions from the all-powerful agricultural cooperatives, which tell them what to plant when, what combination of fertilizers to use, what type of pesticides to spray and how often. This system of directives spit out by computers ends up determining the farmer's daily work schedule. By following these to the letter, the farmer can have the cooperatives pick up the resulting produce and receives payment automatically by electronic transfer to his bank account. So he doesn't have to do anything; everything is handled for him. As long as he takes his money and does as he is told, everything goes smoothly, including his day of work. If by the slightest caprice, one had the ambition to try not applying pesticide; if, upon intentionally skipping a single application of pesticide when the eggplants should be sprayed once every four days, the eggplants take on just a slightly yellowish tinge resulting in 30 percent rejects, one faces more than just a poor harvest. You are told in no uncertain terms, "Look, you've got to stop growing crops." People in the cities have no idea just how bound up in this agricultural cooperatives system farmers are.

A farmer hates nothing more than tests. They just don't do any. If you do something different from everyone else, you are a hobbyist. So it is the experimental stations that run the tests.

I believe that most agricultural testing stations west of Tokyo are paying some attention to natural farming. Many centers have been running studies on my methods for years. They don't call it "Fukuoka-type farming," but they've begun working with my methods—always with scientific modifications of one sort or other, of course. The results of all these tests are beginning to come in.

I am certain that once natural farming wins general approval among agricultural specialists, farmers everywhere will follow suit. At the same time I suspect that, until the results at all these centers are out, any effort on my part to address farmers in general will be in vain. I myself am very well aware of the role played by the testing centers, which is why I have never once volunteered to advise farmers directly. The only place I have given any direct guidance recently was at a convent in Kyūshū, upon invitation.

I also gave some instruction once about twenty years ago for a period of one year on the outskirts of Matsuyama. The head of a local agricultural extension office had gone to learn about direct-seeded, no-till rice cultivation at the agricultural testing center in Hyogo Prefecture. Over there, they told him that the method was being used by a Mr. Fukuoka in his own prefecture. He was very surprised when he heard this and, coming back, asked me to help him out. So I assisted him in carrying out an experimental program of direct-seeded cultivation at several local villages. All the farmers at the time were opposed to this at first, but eventually they agreed to participate.

At the agricultural fair, rice grown with natural farming

won first prize. Second prize went to rice grown by conventional scientific farming methods, while third prize was also copped by a natural farming entry. With the results split both ways, the conclusion reached was that either method of cultivation was acceptable.

But the director was transferred back to prefectural headquarters. With just one specialist remaining, it was impossible to give proper guidance, and calls for advice kept my phone ringing night and day. People would call and ask, for example, "Should I let water into my field? I wasn't planning to, but my wife went ahead and did so. Then my father drained it, saying that my wife shouldn't have done what she did. Now I've got a quarrel on my hands. What are you going to do about it?"

I realized then that unless the prefecture, the agricultural testing centers, and the agricultural specialists and advisors all work together, a cohesive program of agricultural guidance is impossible. So I backed out of the whole affair. I had devoted an entire year to this, but in two or three the effects all but disappeared.

At first my method began catching on quite well in neighboring prefectures. I can even remember thinking that at this rate natural farming would spread rapidly. But the Agricultural Ministry was pushing modern farming with large machines, and strongly encouraged the mechanization of transplanting. So my method, which requires no machinery or fertilizers, was snuffed out just as things were beginning to get underway. That is why I kept my mouth shut for so many years.

Today, however, judging from the speed at which my method of barley cultivation is spreading, I think that rice cultivation by some kind of compromise method between natural and scientific farming might also catch on well. This, combined with recent improvements I've made in my methods—which have been simple but very rigorous—that ensure success even by novices, should also make a difference.

Changing Attitudes Toward Nature ───────

Japan's beautiful countryside has been crafted by farmers over thousands of years. They served nature, growing crops without digging up and destroying the soil. In this way they protected the earth. That is where lies the splendor of the Japanese farmer. But this tradition has not been handed down to young people. There are, I believe, several reasons for this. To begin with, social conditions underwent rapid changes following the war. The thoughts and beliefs concerning work and immersion in nature formerly held by farmers were cast aside. Farmers in the old vein did not have the time to pass on such sentiments to the young.

Another reason was that, before the farming youth of Japan had acquired a feeling of affection for nature, they became dazzled by the vigor of modern farming's rapid development and the splendor of its scientific techniques.

Yet a third reason was that the latent fertility of the earth was such that for a while exploitative farming methods based on modern science yielded economic benefits. Even after decades of unrelenting destruction, the earth has retained such a deep residue of fertility that exploitation has been able to continue up until this day. Despite year after year of persistent assault, the soil continues still to yield rice and barley. Finding that they are able to grow crops as long as they use pesticides and fertilizers, farmers have depended to this day on the goodness of nature, thinking it all right to destroy the earth and profit financially by this.

However, in the West, such an indulgent nature has already disappeared. The land may appear beautiful on the surface, but because the earth is hard and depleted, farmers have no alternative other than to rely on chemical fertilizers. In fact, the energy poured into the land is sometimes greater than the energy recovered in the crops. This is why farmers abroad are financially insecure and unable to turn a profit. As a result, European farmers have to cultivate acreages tens of times as large as those of Japanese farmers, while American farmers must farm a hundred to two hundred times the acreage of Japanese farmers just to make ends meet. On top of this all, regardless of what they do, life for these farmers is just one long struggle—which is only to be expected. Starting off as they do with an awareness that the methods they have used up until now are no good, they are prepared to give natural farming a try since they have no other alternative.

In Japan, even though farmers have killed and maimed the soil with science, they feel that they can still get by for a while with the same methods. Becoming concerned on account of the fracas over food contamination, a few have decided to try natural farming. In Japan, a great number of people have chosen to switch to natural foods because, with the pollution problem, they don't trust normal food anymore. But I feel that approaching things from this angle is a mistake.

Westerners have noticed the defects of the body when it is nourished on an unnatural diet. They have lost faith in the sustenance of life and no longer are able to count on the grace of nature about them. Beginning as they do with the realization that they must live by their own efforts, it is not for them a matter of good or bad. They have nowhere else to go, so they try this. They come to me, saying, "I know that things aren't going to work out with Western agriculture, nature, and this body of mine, so I'm setting out on this new road." It's like the triple jump. Westerners come to my farm after taking that first hop and step; all that remains is for them to make the final jump. But the Japanese who visit me are still lingering at the start of the runway.

The Japanese like to weigh the relative merits of the East and West. "We've followed the Japanese way of doing things up until now," they'll say. "So we ought to try adopting Western methods from now on. What could be better than combining the two?" In contrast, people in the West have fallen into despair and are searching for a totally different road. The Japanese are simply hurrying after Westerners along the same road to failure. Come to think of it, they're passing by the Westerners, who have doubled back and are now moving in the opposite direction.

Before long, Japan's natural beauty too will be completely destroyed and replaced by a false nature. But no one seems aware of this yet. The first sign we have is the epidemic of pine rot. I think one can safely regard the rate of collapse of Japan's pine forests as the rate at which nature is collapsing in this country.

I feel that a way for reviving the ravaged nature of the West exists in the traditional attitude of care and respect with which Japan's farmers have treated the soil. For this very reason, the spirit and soul of the Japanese farmer must be renewed once more. But that soul has been lost; no one is willing any longer to crawl about barefoot in a rice paddy. Here is a short piece I wrote for the *Asahi Shimbun* in 1983.

*   *   *

Reviving the Soul of the Thousand Fields ———————————

The thousand rice fields built up with such toil by our ancestors over thousands of years are a study in miniature of Japan. The mystery of the Orient can be found in the sight of the farmer clinging to a high stone wall plucking weeds. These terraced rice paddies conceal within the world's richest soil the accumulated skill of centuries.

Today, the moon reflected in the individual fields is broken and scattered into a thousand different images by a thousand different eyes. But the moon that hangs clear in the sky is unique. God, nature, and the soul of man, all of which appear to differ, are one. This has always been, and shall always be, the only way for man to live. The only way is to live a life without apparent goal, to reject completely riches and honor and live in concert with nature.

Sorrow should flow away with the moon reflected in the flooded paddy fields. The five elements essential to man (earth, water, fire, wind, and light) are to be found at the hearthside in a thatched cottage; that already is enough. The hearthside is the universe; to own nothing is to never be without. Culture and the light of religious teachings are not to be found in the cities and temples of dazzling splendor.

Poisoned by modern scientific farming based on Western philosophy which separates man from nature, the land in the farming communities of Europe and America has fallen into decline. Both the vegetation and food have become false.

Agriculture is the root of culture. If the method of agriculture is mistaken, this upsets the eating habits and culture, destroying a people.

The crowning work achieved by blending human artifice with nature is the thousand paddy fields. The human spirit and nature (God) that dwells in these thousand fields is in the process of dying throughout Japan. Ancient ruins may perish one day, but the thousand fields must never die.

My ultimate goal has been to establish a way of natural farming that makes no use of tilling, fertilizers, or pesticides, and to put this into practice on the thousand fields under adverse conditions. Today, at last, it seems as if that goal has come into sight.

Once the threesome consisting of the ancient soul of Japan's farmers, Eastern thought, and Eastern methods fit together, then and only then, will a richer, easier, and more relaxed form of agriculture emerge that preserves Japan's nature as in the past. That, in a sense, is our dream.

The World of the Bushman

Natural Farming and the Bushman's Way of Life ———

Back a couple of years ago, some of the major television networks in Japan ran a number of shows and features on the Bushmen of southern Africa. This TV coverage was part of a passing rage, but it set me to thinking about the way the Bushmen live and whether the setting they inhabit is the same nature to which natural farming aspires.

To be honest, I didn't watch many of those programs, but from the little that I did see, the Bushman himself is fine. That dazzlingly bright, cheerful face says it all. I felt as if I knew where it came from. Civilized society may mock and make fun of him, but all that is of no concern to him. This is the face of a natural man able to immerse himself in his own joy.

What bothered me, however, was the environment in which he lives. I saw him digging up some kind of root with a stick. He had to go quite a ways to find that root and also to fetch firewood. That means that nature is scarce there. Why the area in which he lives is so barren I don't know, but if the Bushman really led a primitive life and were to practice natural farming, say, I'm certain that there would be a greater abundance of nature around. His seems to be a natural existence but is nothing of the sort. The nature in the background is just too poor. I would say in fact that this is an unnatural existence. In other words, the balance of the Bushmen and the animals and plants with nature has been destroyed.

To determine what or who caused this destruction will require closer investigation. But this much I know. If what is over there were really a part of African nature, and if this was where the Bushman arose and where natural plants and animals once lived, then one would expect things to be quite different. There would be the wherewithal for a richer, more abundant life.

As for the Bushman himself, I'll grant that he has a wonderfully happy and cheerful face, but I saw no evidence that he really knows nature. It is a mistake to assume that primitive people or people who live a primitive existence are necessarily natural people. Perhaps the face of the Bushman is like that of an infant—a face that does not know nature and has forgotten God. People today have moved away from and forgotten nature. But those primitive people, those Bushmen too, may also not know nature; perhaps they are only living in a place that was abandoned after nature died. Like modern man, they too may not know nature. They are happy perhaps because they do not know that they live in a world abandoned by nature and God. Their's may be the bliss of ignorance.

My ideal vision is to dwell together with God and to live a life in which one genuinely enjoys nature. What is especially important for this is to know what true nature is and to have true nature recover.

The land of the Bushman is not a utopia. He lives a life of hardship and scarcity in the midst of an unnatural environment. This is really only an awkward, scrabbling existence in which he has surrendered himself to the desolation of nature. His is just an impoverished world. No true ecological balance exists here. What precarious balance does exist is in the process of collapsing. At this rate, the Bushman will only continue becoming even more destitute. When there is a true ecological balance, nature tends toward greater abundance, enriching human life. By richer, I mean ample microbial life, rich plant growth, and fertile soil; a lively place where animals multiply and all life abounds. That is how I would expect things to be. But the Bushman, who barely manages to get by, appears to do nothing but accept with resignation his barren environment. This is poverty of the body and soul. I do not glorify poverty.

Nature is fundamentally perfect. Here are to be found the most exalted truths, the highest good, and the greatest riches. Both spiritually and materially, nature is replete with the greatest possible wealth. Nature is a place where flowers bloom and birds sing, a place of verse and song. Here lies everything. It is a paradise where joy and contentment reign. This is the direction in which nature progresses. Nature is never one moment late or tardy. It moves freely and innocently as God wills it.

In a sense, however, nature does not advance or retreat. I may appear to be contradicting myself, but when viewed from a natural perspective that extends beyond space and time, these are both one and the same thing.

Nature is always absolutely perfect; it flows constantly from perfection to perfection. The external form changes with time, but nature itself is immutable and unmoving. There is no superior or inferior in God and nature because the opposing states of perfection and imperfection do not exist. Viewed scientifically, relatively, myopically, nature may appear to move from simple to complex; it may appear to progress and advance from imperfection toward perfection. This, at any rate, is what Darwin's theory of evolution implies. But of course, such is not the true state of nature. At best, this is only its outward form.

I myself was amazed to find, when I took the perspective of natural farming, that the insects in my fields create new varieties of rice. This appears to support the idea that nature creates many things at random, becoming ever more abundant. Take also the acacia tree. Not only does it enrich the soil each year, when its flowers bloom, it provides an almost infinite supply of pollen for the honey bees and it scatters an incredibly large number of seed that surely appear wasteful but provide nourishment for the insects and birds. This too can certainly be seen as evidence that nature moves of its own accord in the direction of thriving abundance.

The soul of nature does not seek decay. Man is invariably

responsible when nature collapses or falls into decline. When heavy rains cause a hillside to collapse, this appears to be natural destruction. But from a macroscopic vantage point, nature does not destroy itself; it merely adopts many different forms.

Whenever true natural destruction appears to have taken place, man has upset the course of nature at some earlier point and created the cause for this ruin. The perishing of the soil and the flourishing of artificial crops spell the destruction of nature. Even if the rice harvest is plentiful, the soil at one's feet has wasted away; insects and frogs no longer inhabit the field and dragonflies no longer fly overhead. When no poem or song remains, nature has died, leaving man to live in material and spiritual destitution.

If nature is truly rich and abundant, then man should be able to live a life of plenty in its embrace. To me, it is the sight of row upon row of horticultural hot houses that reeks of poverty.

According to my calculations, by planting a single black wattle seed each year, one could rebuild a fine mountain hut quite a few times during one's lifetime. If the house were destroyed by a storm or earthquake, this would be an occasion for rejoicing at the opportunity to build a new abode. Which is more enjoyable, I wonder: this or enduring for one's entire life the humiliations the world has to offer while amassing the nest egg necessary to build one's own home where one can at last breathe free?

What I am talking about here definitely is not a return to the past. I suppose one could call it a return to the present. It involves neither attachment to the past nor expectations for the future, but simply living in the nature of today. All that is required is that we surrender ourselves to the current of nature. If one rides the great current of nature, then there no longer is any fast or slow. There is a tempo to nature, and yet there is none. In nature, time exists, and yet it does not. We distinguish between morning and evening, but nature has no night or day. Once people are able merely to awaken when morning comes and sleep when night falls, then time ceases to exist. Even the question of whether nature is abundant or not becomes immaterial.

One must live fully in the present, not being swept off by the current, but directing all one's energy to this instant in . time, like the cormorant fishing for *ayu*. That is a life of true wealth.

Nature is an astounding reality. One must constantly keep in mind that coming into contact with true nature can be an overwhelming experience. This is, after all, a world of inspiration that can justly be called the "Great Spirit."

Touching the Great Spirit

Some time ago, a procession called, I believe, the "March for Survival" set out from Tokyo, made a circuit of Hokkaido, and late, one cold, snowy night at last reached my farm. I noticed a young Native American woman among the group of visitors stretching their hands out to the hearth fire and warming their bodies. "Why have you walked all this way?" I asked.

"I've been searching for myself," she replied.

Without thinking much of it, I said, "But you're right *here*, aren't you?"

She looked up in surprise and stared at me for a while. Then suddenly she cried out, "Oh . . . Great Spirit! Great Spirit!"

That was the first time I had heard these words.

The next day, she said that she had decided not to go on to Kyūshū with the other marchers and was returning to America. Everyone was stunned, but when they saw her radiantly happy face as she clung to me, no one made a move to stop her.

When it came time to part, the sight of her embracing

everyone and shedding large tears moved us all. Later, I dedicated a poem to her in my heart. It was a good poem, but on parting I simply gave her my blessings with everyone else.

I imagine that when she touched something wonderful—perhaps the homeland of her soul or the spirit of God hidden within the bosom of nature—the words of exultation that sprang out unintentionally from her lips were "Great Spirit." This was for me a wonderful day in which I was able to touch the great spirit of the American Indians and sense the breadth and splendor of the soul of Mother Nature.

Agriculture for Tomorrow

Natural Farming Offers a New Future ——————

I have already pointed out that natural farming is not a primitive form of agriculture. It is neither a type of organic farming nor a farming method from the past. Natural farming is a way of farming that transcends past and future. It is, I believe, a way of farming practiced from the days of Gautama, from the days of Gandhi. Only, it has not emerged in any concrete form. There may be people who know the reality of God. But, as in the past, people in general today do not know God or nature and do not recognize that he exists.

I do not find it surprising in the least that there remains nothing concrete which might be called Gandhian farming. It would seem only natural that there be a form of farming which attempts to do as little as is necessary to grow crops for human sustenance. If, by aiming for a truly easy way of farming, people were able to live off the land as easily as the birds pecking for their food, one would expect this to survive as an enjoyable way of natural farming. On the other hand, if raising rice in the thousand paddies were seen as nothing but arduous labor, a life where the heart is split asunder by the many reflected moons in the fields, then this would be a difficult way of farming and would not last. I think that until one can respond affirmatively to the question of whether there can be true joy in cultivating those thousand fields and whether there is any value in the experience of farmers who know nothing else—who are born there, work there, and die there, until one can affirm, "Yes, indeed, there is value there; that is the highest form of existence," the system of values held by both farmers and society in general will have to change entirely and current methods of farming will have to be replaced with easier methods.

This is why natural farming is more a way of farming for the future than from the past. The natural farmer is able to gaze at the moon reflected in the many paddy fields and enjoy a quiet life free from the cares of the world. True human life through natural farming in the family vegetable garden—that is the ideal that I envision.

If the laws were changed, instead of everyone converging on Tokyo and the other big urban centers, the 120 million Japanese could spread out over the 15 million acres of arable land this country has—that's a quarter-acre for each of the 60 million adults. Without using machinery, they could build a house on their plot of land and grow everything they need there— vegetables, fruit, grains. If, in order to create a surrounding shelterbelt, they planted a single black wattle seed or sapling each year, then in ten years time, even without a single drop of petroleum, everyone would have plenty of fuel for the home. Japan is large enough to make this possible.

People will object, "But what about cars? What about this and what about that?" Yet if they were prepared to reject the sort of lifestyle where everyone runs around in cars and were willing instead to enjoy life in a mountain retreat, all the absolute necessities are right there at one's feet. One could enjoy a spiritually elevated life without the least privation or inconvenience. Ideally, this would take the form of small, self-reliant communities. All matters would be taken care of right on the family farm. The home, the community, and the country—all products of the nature in that region—would be fully capable of self-sufficiency and self-reliance. When this happens, then all the people of the world will at last be able to join hands in a position of equality.

Foreigners Are a Determined Lot ——————————

When someone decides to quit his job and start farming
naturally, there is no question that he will be able to grow
enough to live on. But what most people are concerned about
is whether they can earn their living this way. At least, this is
what most concerns Japanese who are considering the move.
They often say that they have a wife and kids and that what
they worry about the most is how they would have the children
educated if they settled in the backwoods somewhere and
began to farm. Of course, educating their children would be
impossible. For the first two or three years, they would be
able to eat, but if they settled in an area so far out in the
country that there is no electricity, then educating the children would probably be out of the question. Most people are
concerned too that without electricity they could not have the
sort of modern lifestyle we have today. That, I think, is probably correct.

But let me mention what one fellow by the name of
Kobayashi did. A graduate of the prestigious Tokyo University,
he arrived at my farm with his young bride and a newborn
infant and stayed here three years. Before leaving, he told me,
"I'm worried about how we're going to educate our boy when
he gets to kindergarten age."

"My God," I said. "You're a bright fellow who graduated
from Tōdai* and you're worrying about whether you'll be able
to educate your kid?" That started him thinking and helped
him make up his mind. They have since settled somewhere
deep in the mountains of Mie Prefecture, cleared a half-acre,
and brought in a calf. When I saw him recently, he told me
that they hadn't even needed a half-acre; a quarter-acre would
have been more than adequate. He is studying Shinto language

This sort of thing is much easier to do while one is still
single. That is why, even with a commune of some sort, if it

* Tokyo University.

is made up of unmarried people, things tend to go fairly well. With married couples, when one is willing and the other is not, this creates complications.

Foreigners are a lot more easygoing about such things. For instance, when a couple starts having an affair, the others around them look the other way. That wouldn't happen in Japan. Here, everyone would start asking whether you plan to get married or not; they'd badger you and show concern. Westerners would just tell you to do as you please. In the West, you shut yourself up in your castle and don't interfere with the affairs of others. People let you alone. In Japan, however, the people around you butt in all the time, something which is hard to put up with. Such a tendency even makes communal living difficult. The fundamental difference in thinking between the two cultures may have something to do with this. The same is true with natural farming too: when a Westerner decides to try it, he goes right ahead and does so. The Japanese prefer to talk things over first with Mom, Dad, and everyone else. They have to get everybody's opinion on the matter before making a decision. But, if anything, all this input leaves them even more confused and unable to act. In the West, you don't ask anyone's opinion. You decide for yourself and see it through. Children are taught to be self-reliant, which is why Westerners have a strong sense of independence. In Japan, children are brought up by the mother and grandmother in the home. Nothing is done without being discussed by everyone. The decision and the doing is always a joint effort.

I've heard, for example, that children in California are allowed to decide for themselves when they reach school age whether they will attend school or not. If they decide not to, then the parents have to teach them at home. Children are brought up so that they can decide on such matters on their own, and parents respect their decisions.

Once Westerners have decided that natural farming is the way to go, whether they can feed themselves or not in this manner becomes of secondary importance. Since one cannot know unless one tries, they put this concern on the back burner and go right ahead and give natural farming a try.

To Westerners, the Japanese appear indecisive. It is hard to tell whether the answer they give is yes or no. They are attentive to every little detail but do not seem to have made a decision. For example, when Westerners say that nuclear weapons are no good, even if they are willing to admit that nuclear energy may be useful depending on the use to which it is put, they are dead set against that too. The Japanese, on the other hand, appear to be a cunning people without strong convictions who weigh the good against the bad and choose whichever offers even the slightest advantage.

In the West, those who think that Western civilization has come to a standstill immediately run off to investigate Eastern culture without even looking back. The Japanese, on the other hand, pick and choose whatever they like from both civilizations, but come to no definite resolution. Knowing that modern scientific farming founded on Western philosophy is responsible for the destruction of nature while Oriental practices protect nature, many farmers in the West have converted to Eastern farming methods without the slightest hesitation. But the Japanese people prefer to carefully measure the merits and demerits of both and choose whatever suits them best at that moment. And they do so only after discussing the matter with everyone they know and taking the longest time to decide. Young people in the West decide on the spot without even talking it over with their parents.

At the moment, given the character of Japanese people today, there seems little chance of natural farming catching on well here. This, in spite of the fact that the traditional cultural climate of the East appears suited to the development of natural people of non-action. To foreigners, the creation of a natural farm is the creation of a utopia. The Japanese think of Eden as a mythical world, but people in the West today view Eden as a utopia vibrant with life; they feel that present society can afford to waste no time in returning there. The priest I met in Holland who dug up his lawn, created a vegetable garden, and found there his Eden illustrates very well what I mean.

There is much talk lately in Japan of protecting the environment and restoring natural vegetation. But when I speak of returning to nature and creating anew the sacred village groves of the past, the only response I get from people is: "For what? Taking hikes in the woods?" The idea of building there a utopia never even occurs to them.

Generally speaking, even when I talk of returning to nature, most people in Japan have no idea where nature is, and are unable to see what is there once they do return. So it is easy to understand why they don't get serious about it. Unable to tell true nature from false nature, they often mistake mere imitation for the real thing. It is no surprise then that they are unable to think of nature as a utopia.

Life by the hearth is what true living is all about. Here one has nothing, but one at the same time has inexhaustible riches. All five elements of which the ancient Greek philosophers spoke are near at hand here. But when this is just an idea in the mind that has not been directly experienced, one senses only the smokiness of the hearthside and feels none of the joy of infinitude.

The hearthside is the universe and the universe is a drunken dream within a crock. Perhaps it should be said that, rather than looking to a world of the sublime, it might be better to begin by the side of the hearth.

Utopia-Building

The structure of Japanese farming communities probably arose naturally from the needs of life itself. For example, the common, five-family neighborhood units may have arisen partly because it takes at least four people to carry a coffin.

Three houses over yonder and two here—that made up the smallest community unit. This grew to ten-family units for the joint management of roads and irrigation canals. The setup of today's farming villages probably had its start with cooperative transplanting and harvesting.

In the larger towns, a different pattern of settlement can be seen. For example, in "temple towns" such as Kyōto, what probably happened was that students and disciples gathered near a monk or priest and resided in a given area, forming a spiritual community. But as married couples arose and others retired from active service, these began building separate wings or independent dwellings. When people started putting up fences around their homes, relations with the neighbors grew distant. I suspect that such areas were encircled with outer moats and strong walls to prevent this and maintain community cohesion. In these temple towns, then, what began as spiritual communities developed into the neighborhood communities we still see today within the cities. I believe that initially such communities were created in an attempt to establish utopias.

A city planner once told me that if Tokyo's one-story houses were all replaced with 20-story apartment buildings and condominiums, this would open up immense spaces where vegetation could be planted to create a pleasant urban living environment. But somehow, I doubt that tall buildings surrounded by an artificial nature could ever become a pleasant and permanent home for man. Man is born and dies on the earth. When he parts from the land, he is no longer able to maintain the stability of the heart.

Many of the young people who have come to my farm have held the earnest hope of becoming true natural people. They have been free and broad-minded enough to be able to reconcile an open family life with communal living. However, nothing of the sort has occurred on my farm. Young, single people have gathered here, lived communally and studied

for periods of one to three years. Once they acquire the confidence to make it on their own, they always take off again,
like young queen bees setting off from the mother hive to
establish hives of their own.

Actually, the question of whether farming under a nuclear
family or extended family arrangement is preferable, and
whether a communal or a cooperative setup is better really
depends—harkening back to the bee analogy—on the flowers
from which the nectar is gathered. If nature is rich and
abundant, then separate family farming is feasible. But where
the natural environment is more severe, it may be necessary
to establish a closely-knit community.

Even more important is that there exist solitary people
even in the community and that, in spite of their solitude,
these individuals have a great, embracing love for society
and humanity. Man cannot be tied down by systems. The
idea is not to protect people with systems, but that in a
paradise of unadulterated nature, no system will any longer
be necessary.

The question of the private or public ownership of land
would hardly be a problem if those who work could be assured of free access to as much land as they are able to cultivate.
Proprietary rights that exist solely to satisfy cravings for
possession make no sense at all. Since no way exists to
measure greed, there is no halting such desires, which serve
only to push the world into disarray. Far preferable would
be the establishment of a free and fluid environment that
satisfies the desire to cultivate rather than to own.

True, Japan is a small, densely populated country 70 percent of which is wooded or mountainous. Yet, although only
30 percent of the land is arable, this still represents an
average of over one acre of farmland per family, certainly
adequate for growing the food a family needs to live. Land
is precious to farmers, so it is easy to see why they have
been unwilling to part with it. But it would be a mistake to

think this the cause for the soaring land prices lately. Farmers, both today and in the past, have generally bought and sold their land at a maximum of 100 bushels (about 66 hundredweight) of rice per quarter-acre. The skyrocketing land prices we are seeing lately are the result of speculation by real estate companies.

Even though this is a small country, there is land enough here to build houses. It is just that areas zoned as residential land are limited. If people in the cities were able to build homes in the hills and mountains and in the rice paddies and fields, then there would be an infinite supply of land. Land designated as "residential," on which houses may be built, must be serviced by a 13-foot road large enough for a fire truck and by a sewer system. The law does not allow houses to be built in mountain forests, meadows, or fields under cultivation, but one may build huts without electricity or *tatami* flooring wherever one pleases.

If people in the cities took to life in the mountains or campaigned to abolish zoning, land prices would probably plummet. Only one law would suffice—a law stating that houses not be clustered together, but built at least 100 yards from each other. Anyone would be free to build a straw-thatched hut or a bamboo dwelling wherever he pleased. An environment where the water comes from a valley spring and human wastes are returned to the soil is the cleanest; it is a place where people can live a free and pleasant life.

Only crowded residential districts have any need for a water supply, sewage facilities, and fire trucks. And no matter how highly developed modern civilized life is, it can never compare with the perfection of a life in harmony with nature.

What do people basically need to live? If a family has a quarter-acre of land on which they grow rice, barley, vegetables, and fruit; if they make their clothes with cotton and the home is surrounded with bamboo, acacia, and other trees, then they have everything they need around the year for food, clothing, shelter, and fuel. There is nothing else they need run after in a car and acquire. The frets and concerns of society over social systems and laws have nothing whatsoever to do with life in a mountain cabin. If one were to build a home in the wild away from such concerns, and restore about one a rich natural area, this would surely become the paradise of natural people of non-action.

In European countries such as Switzerland, Austria, and even Holland—which is said to be the world's most densely populated nation, the moment one steps out of the cities into the country, one finds houses standing alone in the middle of broad pastures and deep forests, the neighboring dwellings only visible far off in the distance. Few of the roads are straight and paved; most are bumpy, winding lanes along which lie quiet, old houses of wood or brick. I wouldn't have been surprised had a crusader dressed in armor or even Don Quixote himself stepped out to greet me. The Europeans who I had thought had such an advanced civilization are still today enjoying life in the mountain huts of the past.

Of course, I too have tried to turn the natural farm into a utopia, but even if I were able to create an ideal village, there would be no one to live there. Children from the cities today are unable to sleep for fear in my mountain huts with torn *shoji*.* They soon become bored with life in my hilltop orchard. Even if they play catch with the mandarin oranges there, they quickly tire of this and ask, "Aren't there any other fruits here?" When I tell them, "There's a persimmon tree over there," they run away, saying, "I can't climb that tree" or "I don't have time to go that far."

And if you tell the parents to bake some sweet potatoes, they are unable even to start a fire in the hearth. The sunken hearth is an automatic fire extinguisher, but even when I explain that the fire can be left alone without fear of the hut catching on fire, they are afraid. So I teach them to arrange

---

* Sliding screens spread with rice paper that serve as windows and doors.

the firewood in the form of the Chinese character for fire (火), and explain that a fire is kindled differently in the winter and summer. When at last the fire gets started, they are incapacitated by the smoke and baking potatoes becomes quite out of the question. Far from being a civilized person, the urban housewife has reverted to a primitive animal afraid of fire.

The housewife accustomed to the pushbutton convenience of city living has no idea how to go about things on the hilltop huts here. The urban dropouts are more likely to feel at home in the orchard. These are often seen as individuals lacking in common sense who take a road different from the ordinary man. But what happens when people with a perverted approach to things come here? In Buddhism, those off the beaten path are regarded as heretics. Like the spectator who has a better view of the action on the playing field than the players themselves, the individual who strays from the way of Buddha and follows a lower road is still aware of himself and can be saved, whereas one who immerses himself in Buddhist teachings loses sight of the Buddha. The scoundrel in a bind is easier to save than the virtuous man who seems to know it all.

Those who think they understand nature are ill at ease on my natural farm. If nothing but heretics gathered together on this farm, perhaps one or two might win the favor of the world.

I have closed my farm to students, however, and for the most part stopped receiving visitors. One reason is that I myself do not have that many years to go. But my decision to seclude myself is not due so much to a desire on my part to retire as to the prospects I now see of the farm becoming self-sustaining without my having to look after it.

Even without anyone running it, the farm should go on improving naturally. I believe that what I have done up until now is to create the opportunity for a farm ruined by scientific farming to return to its original condition. Given

the slightest chance for the earth to regain its former fertile
self, the farm should recover, become self-sustaining, and
transform naturally into an ideal state. Once nature has been
fully restored, the farm will no longer need any tending. My
stewardship will be over.

Here is a pair of haiku that I wrote in anticipation of this:

> About this small hut,
> The cherry and rape blossom
> And roosters crow out.
>
> Flowering radish;
> Chickens scattered here and there;
> Nobody in sight.

An earthly paradise filled with peach and cherry blossoms,
the flowering radish and rape, the barking of a dog and the
clucking of chickens. Picture yourself savoring tea by the
hearth in the middle of this:

> Roughly hewn by hand,
> The hut ridgepoles stand exposed
> By the spring hearth fire.

This certainly seems to be an elegant way of living, does it
not? Maybe this is no paradise at all, but an escape from
reality!?

5

Nature, God, and Man

The Wandering God

Do Not Name the Nameless God ——————————

I think that in a world beyond words, where language is of
no consequence, "God" and "nature" are one and the same.
When I say "nature is God," what I mean is that the essence
of nature and the essence of God are like opposite sides of
the same reality. What appears on the surface is the physical
form of nature; God lies concealed behind nature. Unfortunately, however, when one speaks of "inner and outer" or
"front and back," because people conjure up images of two
relative things they are unable to see nature on the outside
and God on the inside as a single entity. What is one is seen
as two: with mental discrimination comes deepening confusion. According to Buddhist doctrine, discrimination serves to
split apart God and nature. The gods that we speak of—the
Shinto deities, the Christian God—all become one at the
summit. But when the holy men who know this stand on the
summit and speak to us below, different interpretations arise
depending on whether they call what they see the absolute
God of Christianity or the Nyorai* of Buddhism. Differences
of expression are unavoidable, but I believe these all point
to the same thing.

However, the "God" that people speak of is not the true
unified God at all, but diverse gods. I think it is possible to
say (as Christianity does) that "God is absolute. There is only
one God—the God of Christ." However, this God must not
be the deity sitting on the summit that everyone thinks of,
but a God further up in the heavens. I think it fitting and
proper, therefore, to say, as Jesus did, "There is only one God,

* One who has attained Buddhahood.

who is absolute and illuminates the whole world—the Christian
God." Yet the vision that most Christian believers have of
God is an indirect, mental image conceived as they look up
from the foot of the mountain at Christ standing on the
summit. They do not see the God that Christ saw. A clear
distinction must be made between the God of which Christ
speaks and the God to which the people of our world refer.

The Buddhist adherent sees Gautama as a living Buddha.
However, the Buddha of which Gautama spoke was a Nyorai
that beggars the human imagination—an absolute deity.
Because Christ said there is only one God, Christians believed
in one God, giving rise to monotheism. Buddha said that God
dwells in all things so Buddhists understood this to mean that
Buddha assumes many shapes and forms; this multiplicity of
incarnations led to a polytheistic faith. Hence, Christians and
Buddhists believe in different gods. It is not surprising then if,
although Christ's God and Gautama's Buddha are identical,
the gods seen by these two sets of believers differ.

Buddhism says that God dwells in all things while Christianity says that the Holy Ghost dwells in all things. The former
is polytheistic and thus different from monotheistic Christianity. To most people, "mono" and "poly" appear to denote
different things. But when one stands at the summit of the
absolute world, there is no "mono" or "poly." Whether one
calls that rock over there a small pebble or a large stone, it
makes no difference. In the world of language, "mono" and
"poly" are unlike, but in the absolute realm beyond the relative
world, neither "mono" nor "poly" exist. Everything is one
and the same. Distinctions such as large and small, many and
few in the secular world serve only to confound. Even "some"
and "none," too, are identical.

It is true of the Buddha of which Gautama spoke and of
the nothingness taught by Lao Tzu; regardless of who said
what, these all refer to the same thing. Only the names by
which these are called and the manner in which they are
described differ. One other thing that differs is the position of

the individual viewing God. This is basically all it boils down
to. Such differences arise because one can speak only from
one's own particular frame of reference. The Japanese, for
example, know only the language spoken in Japan. They don't
know Western languages. The vision of God seen by Christians
in Western Europe is only a profile seen from the West.
Perhaps Moslems know only the northern profile of God and
the Japanese only the eastern profile. From this inevitably
arise different manners of expression. But while the manner
in which God is called and described differs, everyone is looking at the same entity.

Understanding That Goes Only Three-Quarters of the Way

Suppose that we have a large mountain—take Mt. Fuji for
example—up which people climb. People from the West
climb it from the left side and people from the East climb it
from the right. Others choose to climb it from the center.
Actually, many different paths exist, along each of which
people climb the mountain.

When a drop of rain falls onto the mountain, if this flows
to the left, it becomes Western philosophy. If it flows to the
right, it becomes Eastern philosophy. When seen from the left,
the individual sitting on the summit may appear as the face of
Christ. Seen from the right, he may appear as the Japanese
deities. And seen from the south, perhaps he has the counte-
nance of Gautama.

Yet I believe that there is only one truth—past, present,
and future. And that truth, regardless of what anyone says, is
absolute and unique. Christians may say that there is no god
other than the Christian God. The Buddhist may insist that
Buddha is the Supreme Being. But just as there is only one
truth, there is only one God. Why then does this unique God
appear to have different faces?

Take the Christian cross (+), the Buddhist sign (卍), and
the Shinto symbol (±), which represents a cross planted in
the earth. These and the many other religious marks conceived
by man have something in common. I believe that they all
try to express the effacement of the world of relativity (relative
thought) by doing away with right and left, up and down.

What sense do the words of Christ take on for people
climbing the mountain from below? When they see a cross
before the summit, the mark on the cross and the doctrine
for which it stands appear to be their final destination. Shinto
believers who climb part way up the mountain will see a Shinto
gateway and think this to be the highest deity. Those climbing
from the south will happen upon a Buddhist temple and think
that the Buddha is present within the temple—as if the Buddha
dwelled in Buddhist scriptures.

This is as far as we can go in anything we do—sensing,
arguing, speaking. The summit is beyond our reach; we can
never go more than three-quarters, or perhaps nine-tenths, of
the way to the top. That is the limit to what we can understand. If we could stand on the summit, we would be able to
see God, but although God cannot be seen on the way up the
mountain, we come under the impression that we do indeed
understand God, and even dare to speak of Him. But God
lies in the sky (absolute world) beyond the summit (relative
world). He cannot be described or depicted through writing,
speech, or graphic images.

In America, I met some people of the Jewish faith. I spoke
all night with them about the religion and thought of the
Jews. I found that they have truly wonderful ideas, but in the
end they are extremely rigid and unyielding. In our discussions
of Christianity and Shinto, we agreed nine-tenths of the way
up the mountain, but our views differed when it came to the
summit itself. If the sky above as seen from the summit may
be assumed to be the same, then no matter where one climbs
from the view should be identical at that point. The sky above
the summit belongs to no one.

This is the same as saying that the sky over the West and

the East, over the Japanese and the Americans, is the same
everywhere. Although everything should come together at that
point of emptiness or void, because we only go eighty or
ninety percent of the way, we are unable to do anything but
imagine what the view is like from the summit. And that is
why everything falls apart. We are unable to merge the concepts of God and bring the religions of the world into a single
common unity.

There Is Only One God

To arrive at the conviction that all these gods are one, we
should begin by realizing that while we may be able to look
up at a portion of Christ's face from where we stand at the
base of the mountain, we are not in a position to know the
God experienced personally by Christ. One has to start by
knowing one's own position; where one is right now. No
matter how much you study the teachings of Gautama, you
may come to know something of Gautama, but you cannot
perceive the Buddha that he saw. Even assuming that Jesus
and Gautama indeed preached God, because the words they
used only deepened the confusion of the human mind, these
were of no help in aiding others to understand. Their teachings served only to point out just how erroneous and mis-
guided are the knowledge and actions of man and, in so doing,
to crack the veneer of man's confidence and force him to
reflect. Even a saint is unable to lucidly describe God directly
with words and bring people face to face with Him. We are
only able to guess at the vision of God grasped by Jesus and
Gautama through the words these prophets spoke. If people
were profoundly aware of the fact that what they know is but
a shadow, a false image of God, then they would be unable
to speak and act as if only their God were absolute—the
highest God. Only Jesus is able to say, "There is only one,
absolute God." Only he who himself has seen God can with
full confidence say, as did Gautama, "I am my own Lord

throughout heaven and earth." Muhammad too said, "There is only one God." But that is to be expected. When saints who have reached the summit sit in a circle to speak of God, they do not need spoken words. As for the words themselves used for expressing the God they have seen, because they are looking up from the summit at the same heavens, they would agree even were they to remain silent. However, these saints have no choice but to rely on spoken words to explain the God they have seen to those climbing up the mountain. But these words cannot serve as a means for transmitting this vision for they inevitably give rise to numerous misconceptions.

The lecturing, preaching, and writing of books on God and the Buddha by religious leaders who neither know nor have ever personally experienced God or Buddha only deepens confusion. The greater the number of holy books and scriptures, the stronger these are thought to make the vine along which people climb to approach God and the Buddha. But rather than clinging onto this vine and working their way up it, people read a great many sutras and deepen their knowledge. What this does is to mislead them, causing them to drop lower and lower down the vine. Knowledge does not sublimate people, it ensnares them in confusion. Intellectual discrimination merely fogs our understanding of God. Synthesizing such knowledge may be helpful in obtaining a concept of God, but this only deepens knowledge and corrupts people, preventing them from ever approaching God.

God, or nature itself, lies above. God lies in the direction one walks, one step at a time, by renouncing knowledge and the self. But people choose to go in the opposite direction, trying to possess God by resorting to all means and knowledge at their disposal, by using their own body. Although the true way lies along the road of renunciation, they believe that it lies through possession. They are barking up the wrong tree.

*　　　*　　　*

Take a single planted gingko tree in a shrine wood in Tokyo.* Is this nature or not? Well, because it has been

---

* Excerpted from a taped conversation in Tokyo.

planted in the middle of a big city, it could very well be argued that this is not nature at all. But aside from the question of natural and unnatural, there is no doubt that the tree *is* a tree, so one could also argue that it is nature. Now, when I say "nature is God," I am asked, "Well then, is that tree also God?" Although I am unable to answer that this is a sacred tree which differs from other trees, I have a feeling that it is probably all right to say something like, "God dwells in that tree." People have a difficult time understanding what is meant by "nature is God." One could not say, for example, that the birds in nature are God. But I feel that it is all right to say that those birds know God or at least that God dwells within them. God dwelled at one time in man as well, but man has forgotten God and no longer even notices this. Of course, the birds probably are not aware of it either because they don't think of such things and don't know the word "God." Nonetheless, they appear to live and play together with God. Man does not know God, and loses his way in search of God.

Where Then Is God?

Rather than me trying to answer this, I would like to turn around and ask people what they think God is. People often say they "think they understand," but "thinking" one understands is worlds apart from actually understanding. They may think they understand absolute nothingness (*Mu*) to be nothingness that transcends being and nonbeing, but "absolute nothingness" is a term understood only by those who live in a realm free of mundane desires and attachment; it cannot be understood by those of us who inhabit the secular world. To us, this word is already obsolete. I dread our casual, unknowing use of obsolete words. If I were to ask whether God is present in this—in any—room, I wonder what the answer would be. Those who understand God must be able to give some kind of an answer. What about the water in this glass? Perhaps one would allow to God's presence in the tea prepared

by the great tea master Sen no Rikyū, but what about alcoholic
spirits, drugs, or flush toilets? Failure to give some kind of
definite answer is evidence that one does not know God.

People think that God feels more comfortable in some
places than in others. They are attracted to small birds and
frogs but hate mosquitoes. Yet, unless one transcends the
questions of natural and unnatural, good and bad, God is
not to be found anywhere. We may sense what we feel to be
God's presence in the birds and trees, but this is nothing more
than a mental image, a guess of sorts. We do not understand
God's presence in a bird.

I am worried that when someone has a conceptual understanding of God, he asks himself whether this is God and
whether that is not, and ends up getting lost in the process.
In other words, because he has not intuitively perceived what
God is, he has no choice but to ponder this with his mind.
Instead of grasping the true God, all he has grasped is the
concept of God. That is why he jumps from one view to
another, asking himself whether God can't be explained like
this or described like that.

I think that before people try to understand God, they must
first see that man is not in a position to understand things. If
they try to understand God a little at a time or attempt to
clarify Him from all angles, this only distances God. An
attitude that strives to determine why it is that people cannot
know God takes precedence over attempts to know Him.
When man attempts to know something, his thought and
wisdom collapse into incoherence. I have already mentioned
how people feel as if they understand what going beyond
nothingness (*Mu*) is even though they have never directly
experienced complete detachment or transcendence. But this
is different than actually understanding, so they continue to
wander about lost, shuttling back and forth conceptually
between the absolute (transcendent) world and the relative
world.

I mentioned birds and trees earlier. There are gingko trees

growing at the Kanda Shrine in Tokyo. If I were asked whether a particular gingko tree there is God, I would answer, "Yes, one might call it God." And then what happens? Everyone gets hung up on these words. Mention the Kanda Shrine after that, and an image of a sacred gingko tree immediately forms in people's minds. Some people picture the gingko tree in the same way as the botanist. They think, "Ah, that's a living thing." What about a bowl of tea? The tea is just organic matter, but if you say that it was prepared by the revered tea master Sen no Rikyū, people may hang onto the words "Sen no Rikyū." Perhaps they will think that they have been tricked. They may think that, if this were really Sen no Rikyū's tea, it should be excellent, but this is just coarse, weak-flavored tea. This thought is already on their minds. "That gingko tree and those birds are living creatures," they think. "But this is just an inexpensive tea bowl."

I made use of the word "transcendence" above. It is clear that some people appear to know that God and nature are transcendent existences when in fact they know nothing of the sort. They ask me to speak about the transcendent world of God, but whenever I listen to what they have to say or engage in discussions with them, the level of conversation immediately spins back down to worldly matters. By no means is this transcendent or anything else for that matter. People carry around in their heads the notion of things as being mineral, plant, animal, or whatever. Suppose that someone points to a gingko tree, for example, and asks me, "Is that God?" When I answer that it is, they could immediately answer, "Then this too is God." But they don't. They can't. This shows that they are unable to extricate themselves from the world of ideas that relies on mineral, botanical, and naturalist knowledge. And it shows that they have no understanding even of the word "transendence."

Everyone seems to know what is meant by the words "God dwells in a tree," but I believe that no one really understands what the presence of God in a tree means. For forty-five years

I have walked about and spoken with many different people, but I have never met anyone who understood this. What always happens is that contradictions arise in the course of the conversation, and the person I am talking with presents an answer himself. I myself have never given an answer. There is no way other than for that person to notice the error himself. If you really think that nature is God or that God is present within nature, then the same response must always emerge. Unfortunately, people give different answers at different times and under different circumstances. This is proof that a single unchanging answer has not been given. Even when someone says that God and nature are one, depending on the circumstances, God may become the Buddha and nature may become something unnatural. Even when we speak of nature, we usually have just a picture of nature in contrast with non-nature in our minds. Because we do not really understand that nature which transcends the nature/non-nature dichotomy, our confusion and inconsistencies show up in the things we say.

God Is All Alone

So what I have maintained up until now is that there is no way to define or even describe God and nature. This cannot be done so I do not even try; and I say, quite bluntly, that it is not possible. I have never once tried to define what God is, or what a farm is, for that matter. The only reason I have not tried is because it cannot be done.

Although I am unable to lead people, I tell them, "I know that God and nature are indefinable." Many people have come to live in my hilltop orchard, but I have not had the power to lead them, so I have not tried. I may have told them to "go in this direction" or "return to the bosom of nature." But I never assumed the role of leader and rallied them on with: "Let's go over there," "Let's return to nature," "If we practice natural farming, there should be a road leading that way."

I do not say such things because, even if I wanted to, I cannot. Perhaps I have said: "Everyone has to seek nature for himself," and "Search for nature as you practice natural farming." But never have I even tried telling anyone: "Go there and you will find nature." Nor have I ever said: "Believe me and do as I tell you and someday you will meet God," or "Nature is like this." Perhaps it would be all right to say: "Nature will take you along to God," but since it is not possible to describe nature and God, there is no way of telling in which direction one should go. This would probably only confuse and bewilder. Forty-five years ago, I'd already said, "No means exist for describing nature and God, so I cannot possibly have disciples." That is why it has never been my intention to gather others around me; nor have I ever done so. I have never told anyone, "Follow me. I shall lead you." I knew that although I wanted to have followers, this was not possible, so I never tried. That is why I have never had a single disciple.

Why then do I persist in talking and writing? Because, as much as I protest that I do not wish to speak, I am constantly asked what I think. All I am doing is screaming out in despair. You see, there is a world of difference between sensing that one is able to understand God and actually understanding God. This is like one who is deaf but listens anyway to the birds calling.

Whenever I talk with people, even if they are university professors, I never take into consideration, as I talk, their specialty or their depth of knowledge in science or religion. One thing I am good at is pointing out to people contradictions between what they say and what they do. And I talk about how talking is useless. People think that knowledge is useful, but I believe it serves no useful purpose. I explain how there is no way to bridge the gap between these two viewpoints. I believe that knowledge is of no avail, so I have made no effort to study or become brighter. When it comes to philosophy, I am a total novice. The only knowledge I have is what I've managed to pick up while flipping through books of philosophy at bookstores. Even with Buddhism, I once memorized the "Perfection of Wisdom" (Prajñāpāramitā hrdaya) sutra, but have since half-forgotten it. I have never read a book of sutras. So not only do I not know how to guide people, I know nothing. But I do know immediately from your words and actions where you are mistaken and where you are trying to go. I am able to tell people: "You are striving to move toward God with your mind, but you are not succeeding are you? You've gotten sidetracked, haven't you?"

So I am unable to pull anyone along or to say, "Head west," or "Head east." But I am able to say something to the effect of "Go to a place beyond that." When they hear this, people in general answer, "Yes, I understand what you've told me." But then when I ask them, "So what are you going to do now?" they respond, "I'm going to do this and that. I'd like to try doing such-and-such."

"Hold on a minute now," I say. "Whether you go east, west, south, or north, whatever the direction in which you head, all ways are blocked. Although you tell me you've understood this, where do you say you're headed? And yet, there is no place to go."

Even though they have told me, "I've understood that to abandon myself means to reject my desires and attachments and return to nature," instead of returning to nature they run from west to east saying, "I will protect nature."

People talk of "going" and of "returning," but they do not know from where they are born and in which direction they are headed. They have even forgotten what sort of world existed before they were born. While saying that they don't know the world of the dead because no one has ever died and come back again from the hereafter, they talk as if they do know, saying such things as: "He died, so you ought to erect a grave for him," and "If you pray for his soul, he'll go to paradise." People do things like that because they *think* they understand. If no one has any real idea where we go after death, then there should be nothing we can do. Only those who know what things are like in the hereafter ought to be able to erect graves or shrines or temples. Gautama surely must have understood that such things are unnecessary. It is because he understood that he did not tell his followers to hold a service for him when he died. Why then is it that Buddhist followers today devote so much effort to erecting graves, reading sutras, and building elaborate temples? The existence of graves is evidence that people have not the slightest idea of what attaining Buddhahood is.

If man were able to cross over to the world of the dead and see for himself, he would soon realize whether all this ceremony and devotion is necessary or not. But since he cannot, he has no way of knowing. Since he does not know, he has no need to do anything. But he pretends, nevertheless, that he does know. He understands conceptually. People know that the shrines and temples are all nothing more than idols in a conceptual world, yet they worship these idols and become captive to them.

That is why I want to say that shrines and temples which make a mockery of the gods and Buddha are unnecessary. Everyone believes that Buddha or the deities are present there, so they go along in groups to visit these places. I myself have no shrine or temple to visit and pay homage to. That may be the only true difference between me and others. Everyone has something to which they join their hands in prayer, but even if I wish to pray, I have nothing to pray to. Perhaps the truth is that I am not allowed to pray, that I am not worthy of praying.

The child who does not know east from west, who does not know whether to clap his hands in overt homage or to join his hands in silent prayer, has no reason to visit the temples and shrines to worship. That is why, when I see people joining their hands in prayer before the temples and shrines, I think that these people are here because they still do not understand. Either that or they are spectators of the gods and Buddha who set themselves apart from others.

God and Nature Are One

When we speak of nature and God, our minds are given over
to the world of thought and ideas. When we are absorbed in
conversation, we concentrate our entire efforts on what we
are thinking and saying. Such activities are akin to the meditation time in one of those new religions. This being quite the
opposite of the spirit of personal detachment, God rushes to
get away.

But when one goes into the paddy field and harvests the
rice or barley, for example, there is no time for looking around
at nature or thinking of God. To begin with, the mind hardly
works at all. All the thinking one does, as the wind blows, is
"I'm all sweaty," or "It's hot." That is about all one is able
to think. When you are busy cutting the rice, you have nothing
else on your mind. The mind is empty—a total blank. As the
mind voids and one thinks of nothing, one is able to come
into contact with nature. But the more we think of something
and the more we worry about what to do, the farther we get
from nature. Even inside the house, when your mind is drifting lazily, you can hear the sparrows chirping and the branches
stirring in the wind outside. But when you are thinking, you
hear nothing. Your eyes may be wide open, but they see
nothing. That is why one is more likely to be able to see the
light and hear the sounds of the wind and water by being an
easygoing farmer. But it would be a preposterous mistake to
think that in this way one has approached God to however
small an extent, for in the very act of observing that the wind
is blowing and the birds are singing, one already cuts oneself
off from unity with what one observes. The realization that "I
heard that," the thought of how "that was a great sensation,"
the reflection that "a chilly wind was blowing"—these are all
in the human realm. Looking is not seeing and listening is not

hearing. If God is absent from the heart, one can see and hear nothing. Even if people have hearts, without the heart of God whatever they do will be in vain. Indeed, one could even say that without this all is in vain.

God, Nature, and Man as One

That is why, instead of mentally picturing the trees nearby, one should go over to look at and touch the trees directly themselves. It is better to go on over and look at a tree than to simply try imagining it, and touching a tree directly is preferable to simply looking at it. Thinking that this here would be a good place to lie down, then throwing oneself down and coming under the impression that one has succeeded as a result in returning a little bit to nature only succeeds in braking one's progress. One is unable to go a single step further into nature. Unless one becomes a fool and renounces one's attachments, one cannot advance to the world of the birds.

Those sparrows, that dog, and that gingko tree have no reason to look up to man. On the contrary, man himself must bow down in respect on occasion for it is these others who are in fact superior. They just pretend not to know. Man is able to cut down the gingko tree, kill the sparrows, and put the dog on a leash and drag it around. That is why he thinks he is greater and smarter. But the truth of the matter is that these others—the gingko, the sparrows, and the dog—are laughing at us. Yet mankind, unaware that he is the object of ridicule, persists in believing only that he has caught and controls this thing called nature. Man's arrogance and pride have torn him from the embrace of nature.

If man were to rid himself of this cloak of hubris and become naked, if he were to abandon himself, crawling around on his hands and knees with the dogs and cats, climbing trees and in short acting like children do, he would approach God somewhat.

But no, he does not take such an approach. He forgets these experiences; he loses all memory of childhood when he was able to plunge into and become one with nature. And then, once this has become just a distant memory, he recalls it with wistful fondness. Turning back upon misted memories, people talk of how nice a thing nature is, expressing this in poems and paintings. Having entered into an unnatural world, they long now for the beauty of nature. They are moved by their fond recollections of that distant past and, taking up the brush, lay these idyllic memories down on canvas. This self-expression through abstract paintings is all an absurd mistake. Can one truly come into contact with nature this way? Can one really touch and draw nature? Having forgotten one's childhood, is someone really capable of appreciating, of painting, of poetizing real truth, virtue, and beauty? I don't think so. I think that, for this very reason, we ought to have words capable of praising nature and adoring God. We may praise nature, but we are not qualified to criticize it. It is the height of arrogance to use nature as a springboard to draw, write, sculpt, or engrave the soul of the Buddha into a statue of Buddha. And the haughty pride of trumpeting thereby one's self-consciousness and declaring this to be self-expression is unpardonable.

Learning about Nature Distances One from Nature ——

One young fellow who stayed for a while on my farm told me, on the eve of his departure, "I'm returning to Kyūshū to start my own way of natural farming." He had been with me for a year and now he was saying, "I understand. I've got the hang of it now, so I'm going to set up my own method of farming naturally at the foot of Mt. Aso." As I listened in silence, I thought, "What can he mean by saying "*my* way of natural farming?" There is no X or Y school of natural farming. Although this young fellow came here to shed his ego and return to nature, his own words of self-expression were proof that he had not succeeded in doing so.

There have even been some who came to visit for a day or

two, found natural farming to their liking, and said, "I'm going to return home and set up a natural farming research center." Now here I am at this age and still unable to come up with a good name for this natural farm. I have not the slightest idea how I should put it. To tell the truth, I began calling what I do "natural farming" just for lack of a better name. There never really was any way to name this. Oh, I've tried calling it "Gandhian farming" and "Lao Tzu farming," and I think that any of these serve the purpose more or less. There is no use at this point in getting all worked up over this again and calling it one thing or another. I don't know whether Gandhi practiced this type of farming or not; I only tried calling it that way because I felt that he would have practiced it had he known about it. But I don't know this for sure.

Can a young man who has been here for just a year talk of "my way of natural farming" when I myself have been at it for nearly fifty years with such limited success? It is blatant self-expression, that's what it is. Before he even knew what God or the Buddha was, he sculpted wood and proclaimed himself a "sculptor of Buddhist images." Is it really possible to say, "This is a Buddhist image I sculpted myself. This is a work of self-expression." If a sculptor who understands virtue and truth were to fashion a sculpture, then even a terrible work could be called an image of Buddha. If it were possible to sculpt perfect beauty, then this might perhaps become the Buddha, but perfect beauty must be truth and virtue. And it must be absolute virtue. The question is, can an ordinary human being sculpt something that is absolute truth, virtue, and beauty?

The Death of God

The Japan Broadcasting Corporation (NHK)* has been airing documentaries on ruins along the Silk Road lately. On these programs, I have seen magnificent Buddhist statues and I

---

* National public broadcasting system in Japan.

remember in particular one brilliantly colored statue of Buddha
in a deep recess within a rock grotto. I was overehelmed by
the power and energy of the people long ago who created
these amazing objects in the middle of the desert, and by the
authority, the force, the splendor of Buddhism. They must
have gone through great hardships to create these masterpieces.
This just goes to show the incredible power of religion. But
one thing that concerns me is whether the multitudes of
Buddhist followers who worshipped that statue were helped
by it or hurt. Just how much help, I wonder, were the Buddhist statues in those ruins on the Silk Road in helping man
to approach God or the Buddha.

All I've seen is this documentary footage and photos, but
in the shots of the surrounding terrain, I noticed many large,
weathered tree roots. These are probably the remnants of
large forests. The area was most likely forested when those
ancient ruins and statues were being built. Perhaps people
gathered here and lived in what was an earthly paradise.
I suspect that the large number of people required to make
all the bricks for building those enormous temples lived in
these forests and cut the trees down, destroying the natural
environment. Destroying all that greenery under the pretext
of building Buddhist statues can hardly have been the right
thing to do. When nature is destroyed, so is God. The sculptor
of Buddhist statues eager for self-expression who first kills the
true God, then creates his own abstract stone images of Buddha on the same spot is drastically upsetting things, to say
the least. This is why I say that it is all right if the Silk Road
ruins are destroyed. But nature must, under no circumstances,
be destroyed.

Had those ruins been useful as a means of conveying the
soul of the Buddha and of Gautama, then I do not think the
nature there would have been destroyed. The regions along the
Silk Road that were once virgin forest and green plains,
including Iran and Iraq, have turned into parched deserts.
Man has destroyed the nature that existed there and created

a place from which the true God and Buddha have been driven away. No matter how grand the Buddhist images he fashioned here and how vast the cultures he created, he cannot be proud of what he has done. In the end, the Buddhist statues will be just that and nothing more—Buddhist statues which do not transmit to others the spirit of Gautama. One could say that immense ruins perpetrated an immense mistake. No matter how outstanding the temple or pagoda, it cannot become a place good for understanding God. Instead, it achieves just the opposite; it creates fear of God and distances man from God.

Various musical instruments were found among those ruins. Now I am not totally unaware of the beauty of music. But while music of various sorts was played to enraptured listeners within the vast, ancient temples, the sounds of the birds singing in the green fertile plains of those naturally blessed countries of antiquity in what is today China, Iran, and Iraq vanished. Man infuses the music that flows from nature into his instruments, and people enjoy the music that emerges from those instruments. But is this okay? Instead of lamenting the loss of his ability to listen and appreciate the music in bird calls that contains real truth, virtue, and beauty, people are carried away by the sound of someone playing the piano. But does that piano music contain real truth, virtue, and beauty? When man turns everything topsy-turvy and subscribes to illusions, the inevitable result is a decline in both man and nature. This I believe is the lesson that the Silk Road teaches us.

When I say that the nature along the Silk Road has vanished, it may appear as if I'm talking of botanical nature. But perhaps another way of looking at this is to say that there once existed there a nature teeming with life where God dwelled, and that when this perished all that remained were lifeless statues of Buddha and stone caves. The stone caves and the statues of Buddha carved by man remain, but the moment that the living nature created by God was destroyed, the God created by nature also vanished. Man destroyed the places where the natural God (true image) dwelled and replaced them with a God created by man (virtual image). It is a grave mistake to think that because the nature observed by man is merely a projection of true nature, nothing can be done about it if this perishes, but everything is okay as long as true nature (God) remains. When the projection vanishes, so does the real image. When even the virtual image vanishes from sight, what hope is there that true nature and God remain? They too, of course, vanish. It is this that I fear. The same was true in the Silk Road documentaries that I watched. I saw places there where God is no more. No true people remain, so nature and God have died.

What disturbs me about the abstract and theoretical question of whether God exists or not is that, in posing the question, aren't we assuming that God exists in a world beyond man? Doesn't this reveal an inner sentiment we have that God remains at some point beyond nature? I am a little worried that this could lead to the facile thought that God will still be around even if nature is destroyed. The belief that God exists even without nature or man raises God to such sacred heights that it creates the danger of chasing Him from man and nature. When I say that the real God lies beyond nature and man, I do not mean this in a literal or physical sense. I am referring to the transcending of ideas.

People normally think that this moment in which they read, talk, or sip tea is the world of reality. They believe that the world beyond this is an unreal world, an abstract world. This me as I talk and sip tea here* is a physically and biologically perceived reality. But we are speaking and acting here on the basis of abstract notions held by man. My drinking tea here is not really a natural situation. I've come to this unnatural city Tokyo, had an unnatural meal, and because I am now speaking more than I need to, I've gotten thirsty. And that is why I'm drinking this tea. This action takes its source in abstract notions unique to man and in human emotions. Here we have

---

* These comments were made in a taped interview.

tea, and people are drinking that tea. But this differs from the animals and birds drinking water in nature. From the standpoint of the bird, this human world is unreal. The sparrows drinking water in the park next door—that is a real scene. However, although it may appear as the same world, as a corner of the same city, this scene of people talking and drinking tea here in a special reception room decorated in the Japanese style is an imaginary world. Speaking religiously and philosophically, this is an imaginary world, while the idyllic and apparently unreal world of the bird over there is more realistic. The world within this room where we discuss God and the Buddha as abstractions is a world of drunken dreams. It is nothing more than a world of mental recreation.

People tend to separate God and nature from man. For example, although it is accepted that with the destruction of nature through scientific farming and other human practices, we humans too will ultimately be destroyed and disappear, the question in most people's minds is, won't nature then reestablish itself? Biological and ecological nature existed before the emergence of man and, the assumption goes, will surely remain even after he disappears. When we say that God, the Buddha, and nature are transcendent, we believe implicitly that even though these may be destroyed, they will somehow remain. But this way of thinking is a bit dangerous.

The truth is that, not only will nature not remain, God too will be destroyed. Underlying the thinking that God will not be destroyed is a misconception of what the word "transcendence" means. The belief that God and nature transcend the world we know gives rise to the notion that they will not perish. But this just is not so. When I say, "God does not rescue or save people. He looks the other way. He doesn't care in the least whether man perishes," what I mean is simply that true God and false humanity have nothing to do with each other. The way people describe it, God is in a world beyond; it always sounds as if he were looking down upon us from the heavens. But that is all wrong. It would be accurate to say that when man perishes, God and nature too shall perish. When nature perishes, so does God. After all, nature, God, and man are all part of one whole.

Nothing is more stupid and cowardly than to say that even if the universe disappears or the earth is destroyed by nuclear explosions, nature will remain. Suicide is not acceptable in the eyes of God. People kill themselves saying that it does not matter if they take their own lives. Would the earth be forgiven for committing suicide as it pleased? God is not looking anyway, so it's okay—that is what people think. But they are wrong.

Here lies the collective responsibility of man and God. What reaching God and returning to nature means is allowing the true God to live, nurturing nature, and unearthing the self. I believe that this is man's responsibility and purpose. This is his ultimate goal, for it is there that the wellspring of greatest joy lies. If it were said, in the language of man, that God has nothing to do with man, then it would seem fair to conclude that God is paying no attention, that He need not be worshipped with hands joined in prayer. But to forsake God is to forsake oneself. I suppose that my way of putting it would be that you are free to die if that is what you wish, but don't you dare drag God along with you and kill Him too.

People are often able to sense what I mean when I say that nature and man are one, but they are unable to acknowledge that God too is a part of this unity. God is pushed away and left alone in the conceptual world. If people understood that man and God are one, then they would see that when man dies, God and nature perish as well; everything melts away. In a sense, the reason I've been struggling so hard these forty or fifty years is because I've been unable to explain God.

We must not settle for relations between false nature and man. A direct and genuine relationship must be found between true man and God.

Our world is an abstract, conceptual world. People must learn to discard these concepts one after another until they see at last that only what lies up to the edge of the field is the

real world. One could say that what I have been striving to do
is to restore an unreal world to reality. Everything I do,
including writing books, is aimed at exciting a comprehension
of God, but all I am able to do is to argue as fully as I can
the futility of argument and discussion in bringing about
comprehension. I argue and debate in order to know for myself
just how useless it is to debate over God. One does not need
a camera or tape recorder to approach birds in the field. No
amount of research will help one to approach closer to them.
No matter how much one investigates the bird's heart, the
effort will be wasted. But by dispensing with such investigation, one will begin to understand the feelings of the birds. . . .
All I do is to repeat these words and thoughts over and over
again.

So, rather than drinking tea and coffee here, the idea is to
shed this and shed that. Even though drinking the tea prepared
by a renowned master of the tea ceremony such as Sen no
Rikyū may help one to experience a bit of the ambience of
God, one does not approach closer to God as a result. Once
you realize this and cut yourself off from this, you eventually
become foolish and empty-headed. When the head is empty,
then you become aware suddenly that a bird is singing over
there. You realize that the bird was close to God and you see
also just how far man was from God. That is the only way
there is to understand. No alternative exists but renunciation.

During the close to fifty years that I've been farming, I've
done nothing other than to abandon one useless practice after
another. In general, people look for things they should do.
All sorts of research is conducted even in Buddhism. They
study Buddhist thinkers and priests such as Dōgen and Shinran
and Kōbō Daishi (Kūkai). They study esoteric Buddhism;
they study this and they study that. People think that the more
they study, the closer they will approach to the Buddha, but
this only takes them farther away. My point is this: What
sense is there in investigating such things?

I think that the work called *Sankyō Shiki* which Kūkai

wrote as a young man is enough. In that one thin volume, Kūkai leaves nothing unsaid. That bit about esoteric Buddhism* or whatever was probably added on later. What I have a hard time understanding is why Kūkai bothered to go to China for spiritual training after writing this. Reading *Sankyō Shiki*, it seemed to me that this could only have been written by someone who saw. If so, there should have been no need for him to go and study in China. It seems strange that he would arrive at such comprehension, write this book, then head off later for China. I find it difficult to believe that his real intention in doing so was to collect sutras and build temples.

People talk of esoteric Buddhism as if it were something special, but I think this is something of a heretic road that diverts from the Great Way of Buddhism. Esoteric Buddhism is like sorcery, but involves divination through a type of meditation. It reminds me of occultism and is something I have no patience for. The Great Ways of Buddhism and Christianity do not lie in such teachings as this. They are a more generous, broad-minded world. What need is there for lighting a holy fire to pray, or for divination? Why is it that, after renouncing and renouncing until there is nothing left to renounce, one should have to worship and pray in this fashion? By praying one becomes found. The world of Buddhism is a world that transcends distinctions over this world, the world of the hereafter, and other worlds. The world of religion being the renunciation of all distinctions over this and that world and the transcendence of time and space, what point is there in worrying that this may be the only world there is and fretting about the hereafter? Cause and effect chase after one another and nature changes ceaselessly, but these are of no concern to God and the Buddha.

Guardian deities have come into vogue lately in Japan. In the animal and plant kingdoms, organisms create almost

---

* Kūkai (774–835), known also as Kōbō Daishi, was a Buddhist priest who founded the Shingon Sect of Buddhism and played an important part in the introduction of esoteric Buddhism to Japan.

extravagant numbers of seeds and give birth to numerous offspring, many of which die. This is the way of nature and is not a cause of concern for mothers. Nothing would be more deserving of the punishment of God than were the guardian deities* to bring the dead offspring to life again, then threaten and kill the living mothers for profit. The role of guardian deities and of religion is to release humanity from such spells into a free world. I believe that the ultimate goal of religion is to go beyond time and space, to transcend the world of relativity. Why should something transcendental be concerned over the hereafter. Why should one who lives in reality forsake reality and devote himself to the world of the dead? What I am trying to say is that people should not have time for such things. It is enough to live this day fully, to throw all one's strength into today. There *is* no other way, is there? People look back at the past and lament, and they worry over the uncertain future. Although they have no idea of where they were born from and where they will go after they die, the moment talk turns to their ancestors and they are told that they will suffer in perdition for the sins of the past, they start forking out monetary offerings. The aim of religion is to go beyond this; its purpose is not to question the past and the future. Instead, however, it abandons the present and turns all its attention on the future and the past. The fact is that religion has made this world an inferno. I'd almost like to blurt out, "There are no dead Buddhas!"

Finding the World of Nothing ——————————————

I never lived each day as fully as I did after I quit my post as inspector at Yokohama Customs. I never knew such complete peace and happiness. I lived in a world of great joy. Since then, I have merely lived my life while recalling those blissful days. I found no joy in anything else. No matter what I thought or did, the happiest moments were those when I had

* This is a reference to the Jizō, a Bodhisattva regarded primarily as a savior of children.

nothing at all on my mind. To this day, I have done many things in many places, but these have all been mere diversions.

In Tokyo, I have been treated to Japan's best buckwheat noodles, but to me the coarse fare in my orchard huts is plenty. If I work up an appetite on my hill, then buckwheat grown on that same hill suits me just fine. One needs nothing; it is enough to be able to eat, work, and sleep. Why look at all sorts of famous movies when it is enough to open the window over there and look out at the green world outside? As for music, depending on how you look at things, even without learning any instruments or knowing the notes of the musical scale, one can enjoy the bird calls and the croaking of the frogs; although noisy perhaps, these mix in their own way to form an orchestra of sorts. So, by renouncing absolutely everything and breaking away from the world of ideas, one can discover—how very simple it is—that the world of nothing has everything. A place with nothing at all has in fact everything and enables one to lead the richest life possible.

Farmers from Yamagata Prefecture arrived once while I was sitting in one of my hilltop huts and said, "Why this is the poorest place we've seen, both spiritually and materially." I felt like replying, "Doesn't the highest joy and peace exist in a mountain hut that appears materially and spiritually destitute?" But I said nothing. One cannot know this unless one has experienced it for himself.

Once I sent off a Buddhist priest with the words, "Return and live the life of a farmer in the mountains behind your temple." And I told a Zen priest from the famous temple Eiheiji: "There's no need to sit forever in meditation. You may try to achieve a state of perfect serenity through meditation, but you're doing fantastically if you're able to free your mind entirely even for a short space of one or two minutes. Aren't there people who spend ten or twenty years meditating and yet are unable to shut out all thoughts for even a minute or two? That just goes to show how confused man is and how difficult a mindless state is to achieve. Might it not be easier to achieve detachment by becoming a perfect fool and working here, swinging away with hoe in hand than by practicing Zen? This just might be the quickest path for you."

Practicing natural farming within nature at a place like my farm is not a bad thing at all. But if, even as they live and work in my hilltop orchard for a year or two, they see the neon lights of Matsuyama glittering off in the distance and think, "Gosh, I'd like to go into town, stop by a coffee shop and have myself a cup of coffee again," they might as well be living in town. And yet, when they return to Tokyo, within a few days they are prepared again to return to the mountain as soon as possible. Dōgen* said of these people, "No matter where they are, they are nowhere." That is the tragedy of man. It is a tragedy, but let's face it, given the choice of living in the middle of Tokyo and living somewhere in the country, one is better off in the country because when one is close to nature, one is also close to God.

That is why I say that the farmer is near to God. By "near," I mean that he is where it is easiest to find the bluebird of happiness. Actually, God is right there behind the farmer, but unless one looks back, He recedes off into infinitude. Yet God cannot be perceived either if he is right in front of you because then he is too close.

Children, God, and Nature

While still children, people understand God, but ten or twenty years later they have forgotten him. It is possible to tell them, "There's no need to understand God. You rejected God even though you understood him." Even though children understand God, when they enter kindergarten, the teacher tries as hard as she can to yank this away from them. The eyes of children looking at pictures see true beauty, but the teacher points to the crayons and paint set and teaches them, "These

---

* (1200–1253). Founder of the Sōtō Sect of Zen Buddhism.

are the colors." From that moment on, they lose sight of the true colors. The moment that the grade school teacher tells the children, "This is green and this is yellow," the children are made to believe that the pigments in front of them, rather than the actual colors yellow and green, are the true colors. They think that trees must be drawn with the color "green." The light that flows from the green leaves changes from instant to instant so that one wonders whether to depict the green of these leaves in which true God dwells as red, green, or perhaps yellow. Nature moves with such speed that there is no time to capture and draw it. The leaves sing a song and the drops of dew falling from them play music. The moment that teachers tear down the green of nature that is a single fabric of art and music—a harmonious whole of beauty and truth and virtue, and teach it to their students in bits and pieces, the minds of the children are split asunder. They break time up into teaching slots, saying, "Today we will have a poetry hour, a music hour, and a social studies hour." That is when the eyes, ears, and voices of the children fall apart. The birds seen at first by the children were sacred birds, a harmonious unity of truth, virtue, and beauty. But once the instructor taught the children to look at the birds as biological objects, the moment he showed them how to depict and listen to birds as the subject of paintings and music, the moment that he pointed animals out as objects of moral lessons and taught the children to love the little birds and hate snakes, from that moment, the minds of the children were ripped apart into a thousand pieces. When the mind of a child is dissected, the sacred birds within the child's mind are dissected and cease to be. Children originally, left as they are, have a true eye for beauty and a heart that resonates with true music. They have conscience that is a morality complete in itself without being taught morality; they follow the will of God and never violate the natural order.

But from the moment that children are taught to at school, the birds become mere animals, outsiders. And the self become an existence forsaken by God and the birds. The voices of the birds can no longer be heard and their beauty drawn. By explaining love, adults teach children hate. It is doubtful that the teacher is deliberately aiming to broaden knowledge and wisdom useful in separating man from God and nature, in discriminating between and dissecting all things, and in deepening confusion. The mission of the teacher should be to try to unify and integrate nature, God, and man; to get rid of the knowledge that drove man from Eden. The proper way is to acquire learning directly from nature that requires no formal studies.

The more this world progresses and the more culture advances, the greater the resulting destruction of God and nature and the greater the distance that separates man from these. Today, God and the Buddha appear to be on the verge of extinction. The claim is made that plant life survives yet in the great cities, that culture and civilization are still to be found there, and that these places have also shrines, temples, and churches. But in one sense, one could also say that God is looking the other way; that, indeed, He has run off.

A lot of controversy has been raging over nuclear weapons. Many people think, "My God, mankind can't be so suicidally stupid as to actually drop the Bomb. Surely, if worst comes to worst, God will step in and save us. Let's pray to God." But praying to God will do no good because nowhere does there exist a God who will listen to our prayers. I just said that God is looking the other way, but the fact is that the moment man lost sight of God, there no longer was a God in this world. No matter how often we visit the shrines and temples, we cannot evade war. Our prayers and piety are of no help. Is it not such eager presumptions that "someone" will save us which lead us to war? The more we discuss and argue over peace and war, the more peace eludes us. If I may be allowed to repeat myself, "He who lives by the sword shall die by the sword." This is an eternal truth. Knowledge destroys God, nature, and man.

Nature Creates God

God is called the creator of all nature, but this Great Spirit may also be thought of as lying hidden within Mother Nature as the force that raises and nurtures it. The form of God finds expression in the form of Mother Nature; mental images of the heart of God may be thought of as arising from within nature and being caught by man.

Thus, the breath of God becomes nature, and the heart of nature makes man human. Nature and God were originally united as one; they should never have been distinguished one from the other. Nature, God, and man share one heart and one life (the nature to which I refer here is not the same as that of which the scientists speak, but the true form and essence of nature).

God is at once the conductor of the orchestra played by nature and a cute, little performer born into nature, which is his home.

The reason it is possible to say that the nature which God created, nurtured, and made to ceaselessly change transcends time, space, and human intelligence and is complete and perfect in every world is that all things and all of creation on earth and in the heavens are united constantly in truth, virtue, and beauty to form a world of radiance. And all things in creation change in accordance with the will of heaven, moving always in orderly fashion without error.

Nature is both the creator of man and his greatest teacher. Sensitivity, reason, and understanding true to man all can be manifested only through sympathy with nature. Judgment and criteria for right and wrong, virtue and evil, excellence and mediocrity, beauty and ugliness, love and hate do not hold if man steps off the Great Way pointed out by nature. There

never has been any other true road for man than learning
everything from nature and living dependent upon nature.

A Humanity Forsaken by God ———————————————

Man has turned his back on God and clearly bared an image
of himself as the only organism on earth estranged from
nature.

The story of the original sin in which man's ancestors ate
the fruit of the tree of knowledge and were banished by God
from the Garden of Eden is not just a parable of the ancient
philosophers. Even in these modern times, and for the future
to come, it scathingly points out the innate failings of man.

Man arose on this earth as one child of nature in the course
of natural evolution. After having acquired intelligence, he
has broken away from a nature ruled over by God and begun
to dash off full speed, a heretic pulling his bow at nature.

This humanity that has turned its back on God and lost
sight of Him is unable to grasp the heart of Mother Nature.
Instead, he gives free rein to his desires, distorting and ruining
nature as he pleases. As he deviates thus from nature, he is
digging his own grave.

Philosophically, it is quite clear that the belief that man
knows nature and can use it to erect a uniquely human
paradise of even greater plenty and happiness has been
nothing other than a smug illusion. Man, who is unable to
know even a single thing, who is unaware even of his inability to do anything, is only a knight-errant charging at
windmills.

Incapable of observing for himself the true form of divine
nature, mankind is scientifically captivated by the superficial
shapes and images of nature that are but reflections of human
knowledge. He cannot know the principles at the core of
nature, yet he plays blindly with nature, making of it a toy;
he trifles with it as with a ball, going into transports of
ecstasy.

The joy that comes from the material affluence of modern man, the peace and happiness of enjoying freedom, are merely relative and temporal illusions. Clearly, these will end up as but short-lived reveries.

While knowing that scientific truth is not absolute truth, the scientist ignores this, believing that the search for truth through the human intellect is directly connected to the advancement and well-being of humanity. As a member of the world at large, he is drawn increasingly into a society of false and fictitious prosperity.

The scientist may on occasion have misgivings about the future. But believing human knowledge, thought, and actions to be manifestations of the divine will of God because God created nature and man, he sees the future of humanity as resting in the hands of God. He can see no reason why man should worry because God will surely save humanity from danger.

But divine salvation is not possible for man, who today walks alone after parting with nature and forsaking God. And the day is long since past when humanity could be rescued by a savior such as Christ. We've come to the point where all it will take to destroy the earth is for someone or something—not an atrocious fiend, but a well-meaning scientist or a weak-hearted politician, a soldier, or a malfunctioning robot-man—to press a single button. Unless all of humanity accepts the story of the original sin and reverses its course, mankind will continue rushing onward toward destruction.

The time has come when we must seriously ask ourselves what man is and why, breaking away from nature and taking leave of other organisms, he began walking a road all his own. But the biggest question that will be faced at the start of such an inquiry will be whether it is possible to clearly establish what human knowledge and thought are. It will be the question of what this beast called man knows and what he is capable of knowing.

The scientist thinks he knows what man is. Or at least he

thinks it is possible to know. But he merely believes this can be answered by biological or psychoscientific inquiries. He doesn't realize that a fundamental answer cannot be attained even by examining the history behind the birth of man, anatomically dissecting man as an organism, and exploring the thinking and culture that are outgrowths of his humanness.

This is because, in asking "What is man?" humanity has really been asking "What is this thing called man?" "What is this questioning mind?" These appear to be inquiries into man, but they are not. The real questions are these: Why is it that man saw himself as "man"? When was it that man was first placed in the position of having to wonder what he is, and why?

The fox and the badger know what they are and yet do not wander about in search of themselves. Why is it then that only man becomes lost in this quest for himself? Why is it that, although the insects never feel doubt and self-repugnance at their being insects, only man is incapable of being content? When did this happen? As long as that riddle remains unsolved, it will not be possible to claim that man knows himself. Until that day, the seeds of human doubt will not run out.

The first and greatest problem that man faces is to fundamentally reconsider when and how the basic concepts necessary for shaping human thought, including doubt, were built up. This boils down to the question of when and how man latched onto the notions of space and time that form the basic framework of the human conceptual world.

Man believes that he correctly and accurately understands time and space, but he should know that he does not perceive space and time as they really are.

I would like to stress here once again both that man parted ways with other organisms from the moment he lost sight of the true nature of space and time, and that the basic proposition that humanity is sliding toward collapse begins in the erroneous human notions of space and time.

God Knows Neither Space Nor Time

Misconceptions over Space and Time ─────────

Scientists believe that they know the nature of time and are able to accurately measure and understand dimensions such as space, area, size, distance. The proof, they claim, lies in man's use of knowledge on space and time to successfully launch rockets and put satellites into orbit. Yet, although space and time as perceived by science do constitute the basic forms underlying the notions of human thought, and are fundamental tools for assuring natural scientific truth, these do not point to or make use of the true form of space and time within nature.

The space and time observed and interpreted by man are forever erected upon the notions of human thought and have currency only in the world of man (Kant calls these *a priori* notions). In other words, absolute space-time etched within nature is God's time; it has no form or shape and cannot be expressed concretely. If differs entirely from the time used by man.

The time known to an insect is not the same as the temporal dimension we divide up neatly into fixed units. To an insect, a moment is an infinite stretch of time. No matter how small the space occupied by a single tree or herb, these inhabit a vast, boundless space. Even should the scientists unable to perceive this succeed after much trouble in flying spaceships out into the cosmos, their feat will only spread wider the clouds of doubt and suspicion, resulting instead in the enclosure of humanity within a smaller universe. Yet they remain unaware of this. They have not noticed that a boundless universe and infinite time are to be found by the side of the hearth. Instead, they go off in search of a place to live in peace, becoming nothing more than wanderers through the universe.

It should have been pointed out that although short and long may exist in natural time, in the eyes of God, there is no fast or slow; that perhaps large and small are manifest in natural space, but wide or narrow have no place there. Large and small, many and few—these are but delusions of man. Because the space and time perceived by man are not absolute space and time but merely scientific notions of space and time, their value changes constantly, making them something on which man cannot rely.

Man must realize that the speed and range of the jet planes so important to him are but a momentary flash of light as seen from cosmic time measured in billions of light-years, and that the momentary flash of a hermit's crazed dream rivals in duration billions of light-years.

What is it, after all, that man has called space and time? What is it that he has used and over which he has wept with tears of joy and sorrow? To man, the decades of life behind one are less precious than this day today, and they cannot be exchanged for even one day tomorrow. What this means is that man does not grasp and control space and time at all. He has merely been tossed about by ideas of relative space and time and made to dance on the illusory stage of empty space and time. Basically speaking, the notions of space and time are not elements that play a direct role in the true joy and happiness of man. They serve instead only to shackle and torment man.

What science has done is to categorize time, to break it up into the past, present, and future. It has created calenders and clocks, and by promoting the notions of short and long, early and late, near and far, it has distanced man from eternal time. Man has tied himself down at one position within boundless space, and there inserted such frameworks as large and small, broad and narrow, surfaces and solids. Knowing extent, man has worried over large and small, many and few; knowing breadth and quality, he has had to face the torment of growing desires.

Man has not known and acquired space and time. Instead, he has become busy within a life fettered by a jaundiced view of space and time, and lost these as a result.

The only way to know true space and time is to live within abolute time that transcends the space-time continuum.

The Last Road Left for Humanity ————————————

Convinced that, because he has begun to fly out into the broad reaches of outer space with the forces of space science, mankind will be able to seize hold of macrocosmic space-time, and believing that biology shows how the root of life lies within tiny cells, people have started dedicating themselves entirely to such inquiries.

From a point beyond the space-time continuum, however, there is no "macro" or "micro." The root of life does not lie within cells or nucleic acid. Nor is it matter that emerges from within stars in the cosmos and is absorbed by black holes.

True life is not something that can be found within the natural world that is the object of the natural scientist's investigations. Rather, it lies in a world that transcends the notion of human life and death, and the matter and physical bodies in which this notion takes its start. What must take precedence in man's search for life is the renunciation of his intellectual notions of life based upon which he distinguishes what is living from what is dead.

Can biologists actually believe that by studying the evolution and development of living things they will be able to determine the structure of the universe and throw light upon how life arose? Yet, instead of studying life itself, they are only chasing after its footsteps and shadows.

An inquiry into life must consist of knowing what sort of drama the life created by God has acted out on the stage of nature; it must delve into how this life has made merry, frolicked about, and received the joy of being alive. Failing to turn an ear to the music played by the orchestra of God and feeling not the slightest wonder at the beauty of God dancing on a magnificent stage, the biologists merely conduct studies on the stage equipment and the broken and discarded instruments lying about, contenting themselves with just a fleeting view of life.

The value of life's workings cannot be measured by the yardsticks of time and extent. The joy and happiness that arise from life cannot be grasped by the notions of space and time.

If God is the true essence of life, then the biologists who think they are studying life are really only dissecting the ruins, the vessel, of life.

Genetic scientists today know that the nucleic acids DNA and RNA within the cell nucleus are the means by which genetic information is transmitted, and they believe that by elucidating this mechanism man can solve the riddle of life and grasp its true nature. Yet, rather than taking part in the solution of this riddle as they think, scientists are only helping to obstruct the free play of life and to disturb natural life. DNA is not an apparatus for creating and launching information. It may be thought of rather as just a temporary relay station that catches life information from God and transmits it to the next information system.

Man claims that the true essence of electricity within nature became known when scientists discovered electricity and how to transmit it by wire. In the same way, when he discovers the means by which genes are transmitted, develops ways for recombining genes at will, and becomes able to create various new and unusual organisms, man will probably declare that he has become capable of regulating the creation of life as he wishes. When man begins to incorporate monkey genes within human genes and mouse genes within the monkey, when he is able to create a beast that is half-man and half-monkey or

transfer the personality of a mouse to a monkey, he will
certainly have learned how to trifle with the life of organisms,
but this will not mean that he understands the true life of
man or the monkey.

The joy of human life means nothing to the great apes;
man cannot recognize the gorilla or chimpanzee that bear
such a close resemblance to him as his children; and would
the monkey be pleased to be endowed with the squeaky voice
of a mouse? Scientists are committing a ludicrous mistake
over the true meaning and worth of life.

The value of life lies not in its possession by man. Life's
true worth becomes manifest only when that life is allowed to
live and lives dearly. The only concern is that life be lived
fully and completely. Just as the worth of man has nothing to
do with his having two, six, or eight arms and legs, conten-
tions that the length of a life or the presence or absence of
genetic mutations have a direct bearing on the dignity of
human life are merely the dogmatic views of scientists.

The scientist has awaited with such anticipation the development of recombinant DNA techniques because he believes
that the quality of human traits is controlled by the quality
of man's genes. But that has exactly the same level of significance as someone applying makeup to alter one's complexion,
or undergoing cosmetic or corrective surgery to change the
shape of one's nose or the length of one's leg.

There are no distinctions between better and worst in
natural life itself. Judgments over the excellence of human
traits and on virtue and evil are made merely from the
standpoint of man. As a result, all human notions end up as
forgeries and empty fabrications. Man's manipulation of
genes differs in no way from a monkey that amuses itself by
imitating man and applying cosmetics to itself.

Hence, no matter how new and outstanding a form of life
the scientist thinks he has created, that creation is just a com-
placent work of man and cannot possibly become a superior
form of life that is universally viable within nature. In other

words, scientists have not managed to even brush up lightly against the essence of life.

Although the scientist appears to be free to play with the biological phenomena of life and indulge in reveries, it is quite clear that this arbitrary and perverted view of life shall provoke untoward changes and invite chaos in the realm of natural life. The emergence of à single abnormal new microorganism could directly endanger the entire human race.

People believe that man has the ability to distinguish between what is normal and abnormal. But just as even the greatest physician is unable to know what is truly normal and what diseased, man has not been given the capacity to correctly distinguish between true nature and what is unnatural. This problem belongs to a realm beyond the domains of medicine and natural science.

Weapons are destroyed by weapons, and knowledge by knowledge. When man, unaware of his own ignorance, takes pride in his knowledge and swells up with conceit over his ability to unlock the secrets of all things and to freely control nature and life with his intelligence, he will destroy himself with that very intelligence.

Although I am but a stupid, inarticulate farmer, I am certain that the following is correct:

> We need only act naturally in accordance with nature. There is no other way but to search with full heart and soul for the true nature-God and for life. That is the last road left to humanity.

I am returning to nature through natural farming. This is the only road there was.

A Farmer's Ode

How many years have I wandered through this world
   where nothing remains the same in search of something certain?
At the end of a long, dark road lies a mountaintop
   cloister.
Up the winding path through a forest of darkened
   cypress I tread.
A nun comes out to greet me with a warm smile,
   lantern in hand.
Now I know where ends the wanderings of my heart.
Here lies the homeland of my soul.

The morning sun rises up over the cypress wood.
As early mists burn off, the fertile plains of Kyūshū
   spread out before me.
Reflecting the quiet passage of time, the Imari River
   flows without limit into the western sea.
The cloudless, azure sky ticks out moments of eternity.
Now I know how vast is the heart of God.
Ah, the Lord's love; how precious the hand of Crea-
   tion.

The heavens being the seat of God,
   those who till the earth beneath are blessed.
(Come let us joyfully serve the Lord)
The singing of the birds in the fields glorifies God.
The rape blossoms speak of Him.
The spring that wells up from the earth murmurs
   eternal truths.
This day . . . a life infinite.

As the last ray of the setting sun touch the chapel
   crucifix.
The evening bell announces the end of a day's labor.
I thank God for his blessings.
(Now I may rest quietly)

Seeding a Real
Green Revolution

We Must Stop
the Advance of the Deserts

I've noticed from TV programs on China and Korea that
mountains in those countries are not covered with green
forests as they are in Japan. Several South Korean officials
in Japan for a cabinet-level meeting once paid me a visit on
their way home from Tokyo. After I talked to them about
acacias for a while, they took some mimosa seeds (the mimosa
is a type of acacia) back with them, saying that they would
plant these on Korea's denuded mountain slopes.

I've heard that vegetation has reemerged lately on the
mountains in Korea, but it is clear from televised scenes of
the region about China's Great Wall and the Silk Road that
these places are so far gone that no trees remain standing at
all. In Africa too, children and grandchildren travel far from
home, going to great trouble to find trees to cut. But they go
right on cutting what still stands. This explains why areas
which were once dense forest have turned into bleak, barren
hills.

Take the case of Brazil. Ever since Japanese trading companies began to cut down the Amazonian rain forests to grow
rice, the forests have given way to desert. In fact, Japanese
settlers are beginning to worry about this and a few have even
come to me for advice, asking whether it would be possible to
practice my methods over there. The dense forests of Thailand
and other parts of Southeast Asia, regarded until recently as
a vast treasury of wood, have been razed, leaving behind a
devastated land. The global decline in vegetation has advanced
to such a degree that tree-planting campaigns are clearly no
longer enough.

So what steps should be taken to curb this trend? I believe that serious thought must be given to effective ways of stopping the advance of deserts. Current action consists predominantly of civil engineering projects such as drawing water or building dams.

I believe that the problem we face concerns the means that should be taken to protect vegetation. Having people in the towns and cities plant trees and providing loans and other aid for the construction of dams and irrigation systems by afflicted countries addresses the results rather than the causes of the problem. Although I don't doubt that this too may be necessary, the most pressing need right now is to find a fundamental way to curb desertification. I believe that we must start by bringing to light the real reasons why these lands have begun turning into deserts, and cut off this evil at its root.

I've had several places, such as the Zen center at Green Gulch, try seeing if they can't find some clue to a way of stopping the deserts with natural farming. Several years ago, I received a letter from the Zen center telling me that the cryptomeria seeds I had sent to the head of the center had taken well and that several hundred saplings measuring a yard or two in height were growing on the surrounding hills. I had sent the seeds after my return from the United States in 1979, believing that cryptomeria would take better than the enormous redwoods I had seen at the Glacier Forest with their shallow roots.

The head of the center died in 1983. Before dying, he told his followers: "Think of those seeds from Mr. Fukuoka as his soul. Plant them carefully." When I visited the center in 1979, I had suggested that the foot of the surrounding barren hills be planted with cryptomeria, the hillsides with Japanese cypresses, and the summits with Japanese pines. These, I said, should be interspersed with green manure trees and fruit trees of all kinds. His disciples appear to have successfully carried out the revegetation scheme that I painted for them.

I have also received good news from the community of young people trying to survive with natural farming in the arid region of the Upper lake hills. They report that they have managed to grow *daikon* radish, cucumbers, squash, tomatoes, and other vegetables. But I imagine that they still have a long way to go before their efforts can be called a true success. Right now, they are barely able to grow enough food to live on, so they are not yet in any position to restore vegetation on the land around them as well.

I believe that, no matter how much we deplore the loss of vegetation and campaign to protect and restore our green lands, nothing can begin until we have a precise idea of how to thwart the advance of the deserts and revive the land.

Countries with deserts must surely be doing research on ways to contain the growth of these deserts, but faith in standard approaches that rely on large public works such as the construction of dams and irrigation networks appears to be on the wane. At least that is the impression I got when an officer active in the United Nation Environment Programme's Plan of Action to Combat Desertification invited me, together some Japanese farmers and other related individuals, to discuss the problem. I think that he was trying to find something in the basic ideas underlying natural farming that could be of value in fighting desertification. Before we try to resurrect nature, we must first determine clearly why nature perished in the first place. We must learn how *not to kill* rather than how to *allow to live*. Constant efforts to protect nature without removing the causes of destruction are pointless.

I have already mentioned how I learned in California that, although science teaches us that rain falls from the heavens, in a metaphysical sense it wells up from the ground. To revive the vegetation on the land, we must remember that grasses do not grow because we water them; if grasses and trees are allowed to grow, these transpire water vapor and clouds form overhead.

The primary reason why grass does not grow has nothing to do with the lack or presence of water. Man is the reason. I believe that man's input of knowledge on the land is the root cause of calamity. For example, he plows the land and burns the vegetation. This kills the grasses growing on the surface by burying or burning them. What happens without seeds to restore the vegetation? When the land is stripped of even one type of vegetation, this upsets the ecosystem and the effects spread far and wide without limit. A small event triggers a mishap in the biosphere that leads to chaos.

Thousands of years of slash-and-burn agriculture, and the widespread baking of earth over the past few millennia to make bricks for homes, temples, and other structures have had an enormous impact on the biosphere. Today, as in the past, it is clear that the grazing of livestock such as cattle, horses, and sheep is transforming the vegetation that grows in the soil and destroying the land.

For some time now I have lived a secluded and solitary life doing little more than simply observing the growth and changes in the fruit trees in my hilltop orchard. When I was reluctant to pull up the semi-wild *daikon* at the foot of these trees, everyone in my family chided me. They thought I was being stingy.

The road divides in two, depending on whether or not people carry away a single plant, a single blade of grass or piece of straw. One way leads to scientific agriculture that depletes the soil, and the other to natural farming that revives the soil.

My experience, after years of observation on my natural farm, has taught me that allowing any animals larger than chickens—such as cows, horses, goats, or sheep—to graze freely on a farm of up to five acres or so depletes the fertility of the soil. Allowing too many chickens to range over the land also is harmful. Strictly speaking, even living self-sufficiently in a small mountain hut on the land has a negative impact, no matter how religiously all human wastes and

ashes from the hearth are returned to the soil. The quickest
way to enrich land is to scatter green manure and vegetable
seeds beneath the fruit trees and leave the place uninhabited.
All that ever had to be done was to interplant green manure
trees among the fruit trees and scatter green manure plants
and vegetable seeds just once. Man and large domestic
animals are enemies of the land.

One summer a couple of years ago a professor from the
University of California came to Japan in search of the
natural enemies of citrus pests. He visited my farm with three
Japanese people: an official from the Agriculture Ministry, a
specialist from an agricultural experiment station, and a
university professor. I suggested that they search among the
trees in my orchard. Within less than ten minutes, on the
second tree they inspected, they found a chalcid fly that is
the natural enemy of soft brown scale, a major insect pest in
America. My visitors collected samples of the fly and left
delighted. Before leaving, the American said, "With the help
of these flies, we might be able to grow oranges in California
without spraying pesticides." All those years of natural farming without pesticides turned out to be of service in an unexpected way.

The fact that this chalcid fly exists in my orchard means
that the orchard has returned at last to a natural state and is
starting to be self-supporting. I left the orchard pretty much
to itself and pesticide-free for several decades. At one time
it even appeared as if it would be ruined, but the soil grew
richer and the cover of vegetation gradually became denser.
As nature began to reestablish itself here, the fruit trees and
vegetables started to grow without fertilizer. In the absence of
pesticides, various types of natural predators appeared and
the trees began to yield fine-looking fruit. In fact, my natural
orchard has been producing better fruit than the neighboring
orchard, which is sprayed with pesticides.

Within this process of restoring nature lies the secret to a

means for combating desertification. Let me explain. Land
worked by farmers and grazed over by livestock weakens;
the variety of animals and plants it supports declines. In a
"do-nothing" nature without aim or purpose, however,
organisms are able to thrive vigorously and harmoniously.
This is as it should be, and yet is it not fascinating? But
when man comes in and cuts the grasses, these no longer
scatter seed; and because he eats and carries out vegetables
and fruit from the farm, seeds from these do not fall to the
soil.

The starting point appears to be whether the seeds increase
in number or decrease. This made me realize once again that
my approach in searching for a way to halt the spread of the
desert in California—I had begun by sowing the seeds of
various types of vegetation—was not in error. What I did was
to mix together the seeds of many different vegetables, grains,
and green manure plants, and scatter the mixture over the
land. This must certainly appear to be a disorderly and highly
wasteful method. The scientist no doubt thinks that waste
leads to failure, but I believe just the reverse. When you experiment with nature, it soon becomes apparent that eight or
nine times out of ten, things don't work out as you expect.
If you fail completely, that means that your expectations are
out of line with reality. The experience teaches you something
totally unexpected, something new and important. That is
why when I encountered the sort of total failure that had
those around me rocking with laughter, far from being
dismayed, I was able to chuckle to myself and enjoy it.

I have put very little of what could be called serious effort
into any research. Instead, I have tried as best I could not
to do anything and to watch closely how I fail. All I have
really done is to sow seed.

Everything Begins by Sowing Seed ──────────

Natural farming starts with the question of when to sow what seeds. With rice and barley, too, the first consideration is on what day to sow the seed. This may seem an easy matter, but the timing is critical. Take the case of barley, for example. Although the planting season is said to extend from September to November, the best time to sow is a period of only about three days. The best period during which to harvest the barley is also only about three days long. A week too early cuts the yield by ten percent. This is still better than being too late, however; if the harvest is a week too late, the barley overripens and the heads of grain bend over, resulting in a twenty percent loss in yield. The difference is that big. The reason that Japanese farmers rush about so feverishly during the rice-seeding and barley-harvesting period in the spring is that they dare not let the best time for these activities slip by.

However, if you leave things up to nature and leisurely watch the rice ripen and the seed fall naturally to the ground, it becomes apparent that the natural process of seeding takes place over a long period stretching from winter through spring. Knowing this, one can no longer say for sure when to seed the rice in natural farming. The rice could be sown on any of a hundred days. It could be sown at the end of the year, in January, or even in March, April, or May. This gives a sowing period of nearly a half-year during which almost any time would seem to be okay. But if you intend to grow the world's best rice, then winter-sowing gives the highest yields. The only drawback is that winter-sown cultivation presents numerous difficulties. Such rice is the most difficult to grow and requires the most rigorous methods. Three days can change everything, five days is the limit, and ten days late is just too late. That is the way I see it.

Vegetables such as *daikon* will grow regardless of when you seed them. But if one aims to produce better *daikon* than possible with scientific methods, one will have to plant at the right time and place for nature, which is more finicky than scientific farming. I may talk of sowing at the right time and place in natural farming, but neither I nor anyone else is capable of saying precisely when and where that is. This varies with the time, place, and year. However, nature solves the problem of time by sowing a large quantity of seed when most suitable. I continuous-crop my vegetables each year by letting the plants flower and allowing the large quantities of seed to fall to the ground naturally.

The seeds that fall to the ground below a single four-year-old black wattle tree, for instance, number in the hundreds of thousands, and even millions. But very often no more than one or two saplings will grow from these. Only very few of those millions of seeds survive. The chances for survival are perhaps one in ten thousand, or one in a hundred thousand. The rest simply perish. To people, this seems a great loss. There is nothing as apparently wasteful as nature. The acacia seeds disappear—eaten by birds, stashed away by mice, carried off by ants. Yet, what appears to us as waste is far from being waste in the eyes of nature. The survival of just a few seedlings out of all those seeds is simply part of the natural ecological process. Nature has left an appropriate number of descendants at the most suitable time.

The upshot of all this is that we can argue about natural balance and imbalance and talk of the sowing season, but when we look at the entire picture, the natural world is so impossibly subtle and complex that it is totally beyond man's ability to comprehend or do anything. With a bit of investigation, we might learn that there are only three ideal days for seeding, and that, given this and that condition, we should use such-and-such a combination and do things in just such a way. However, strictly speaking, we can't even get to square one. It would be better instead to just get the job done without worrying so much over details, and spend the rest of the time napping. Such matters simply cannot be addressed with scientific and analytic knowledge. No attempt should be made to manage nature.

From the standpoint of the acacia tree, it is enough if one or two seeds survive to become saplings. I have no idea how many acacias should grow on a quarter-acre. The fact that the vast majority of the seeds become food for insects and small animals is no problem at all. Nature, seen superficially, is easy-going and generous. That is why "do-nothing" natural farming is possible. On the other hand, the deeper we pursue our analytic study of nature, the more we realize nature to be an advanced organic entity that cannot be comprehended with scientific knowledge. If we intended in such a way to grasp the essence of nature and to practice a natural way of farming that makes use of this, the result would be an incredibly difficult and rigorous method of farming. These are the two differing sides of natural farming. In any case, natural farming begins with the skillful sowing of seed; but the intellect cannot discern what is skillful and what is not.

Can Natural Farming Stop the Deserts?

I have already noted my astonishment at the desert-like character of California and briefly mentioned the seeding tests I ran there. The scarcity of green vegetation in California is a consequence of the many people and domestic animals that live there. What happens is that a large number of people choosy about the food they eat gather to live in one place. They cut down trees, construct buildings, erect temples and churches. All this causes the vegetation to recede. The same thing has happened in Japan. It all started with cultural development in the southern part of Honshū. More than a thousand years ago, provincial temples were erected everywhere. This apparently began to deplete the mountain forests in that region and on the neighboring island of Shikoku. The very fact that the land becomes increasingly poor as one approaches an area where a cultural flowering such as this took place attests to the part played by man in that destruction.

It is quite clear that in many places of the world trees on the hills and mountains have been burned down through slash-and-burn farming, and the vegetative cover has dwindled because large domestic animals such as cattle, horses, and sheep graze the land or did so in the past. Knowing that in Africa, Iran, and Iraq, vast areas that once flourished are now arid and unproductive strongly suggests that these turned to desert through injury inflicted by man.

When man cuts down trees and his livestock feeds on the grasses, the varieties of plant life invariably dwindle. This simplified vegetative cover is easily destroyed. If the land is taken over by yellow grasses such as foxtail, the increased reflection of sunlight raises the temperature of the ground. This in turn upsets weather patterns and causes the complete evaporation of ground moisture, transforming the land into a desert. Such is the order in which events occur. The land does not become a desert because the water disappears. There are fundamental conditions preceding and underlying the disappearance of that water; namely, the perishing of trees and other plants. Hence, a proper and effective approach to combating desert encroachment would appear to begin with the restoration of vegetation native to the affected area.

Up until now, however, since everyone has assumed that the vegetation disappears because of a lack of water, whenever plans have been laid for revegetating an arid region, the first thing done has been to draw in water. Dams are built and irrigation systems constructed. Not only is this approach only marginally effective, as countless examples in Egypt and elsewhere demonstrate, such efforts often end in failure on account of salt buildup in the irrigated fields.

What then should be done? Well, the very first thing would be to cover the ground with a mantle of vegetation. Anything will do. If I were asked for advice on reclaiming the deserts of Iran and Iraq, I would suggest that first a large-scale test be conducted in which the seeds of many different kinds of plants are collected, including those native to Iran and Iraq, and these seeds scattered once over as wide an area as possible just before the rainy season.

This may seem like a brash and foolish method, but it can pull the land out of a hopeless state. I would like to stress that when something like this is attempted, it must be done over a wide area. When the seeds are sown over a broad stretch of land, even if 99 percent fail, some of the seeds will take somewhere. This germination will give us precious clues. Also, seeding should be continued with the full expectation that this will fail again the following year and the year after that. Even with 99 percent failure, anything that shows even a one percent survival rate should be seeded again the second and third year. Start out with a three-year plan, expecting to fail for about three years, and just concentrate on sowing seed over the entire area, as far as the eye can see. I believe that this is the only way to go. Then, and this may seem like groping in the dark, search for the real nature; begin by determining what the nature of ancient Iran, and Italy, and Holland was like. First raise the plants that will grow in the desert, then try gradually increasing the variety and quantity of vegetation. One must ask what will grow *now* in that ground, then find the answer to this question by sowing many different types of seeds.

This is an offering by man to nature. In sowing the seed, man is not growing the plants; he is supplying the materials with which nature can teach him. In a sense, it is an offering to the deities. Man is having the god of the land feast on what he likes. If the god likes what he eats, he teaches man: "*This* is good *here.*"

Relying on his own knowledge, the scientist decides: "such-and-such seems to work well here, so it will probably be profitable," and so he begins. The first thing he does is to study the physical properties of the soil and conduct soil and fertilizer tests. Based on his results, he tells the farmer: "This soil is poor in phosphates, so you ought to grow peanuts." But my method is different.

No matter where one goes, the nature there today is probably not the same as in the past. The nature we have today is unnatural; it is total absurdity, and the best way to begin when dealing with absurdity is to use an absurd method. Start from scratch, sowing anything at all. What you are doing is sounding out the earth, asking it questions. Do this, and the earth will answer back. Such an approach may seem reckless and imprudent, but even a poor shot with a gun will eventually hit the target if he tries long enough. Stinginess, on the other hand, will only be rewarded with silence. The gods will not respond to a frugal offering.

If you summon up your courage and sow your entire collection of seeds over a wide area, a response will be forthcoming. Science may appear to serve a purpose, but the secrets of the almighty nature must be seen through God's eyes to be understood. God will at times give an astounding and unexpected answer. When that happens, one mimics this the second year. And then, in the third year, one sets up a plan. That is the order in which to proceed.

The same approach and methods for converting fields cultivated with scientific techniques to natural farms can be used for changing a desert land into a green plain. In that sense, my natural farm provides extremely fascinating food for thought. I hope to be able to give a detailed report on this at a later date.

What Will Grow in the Desert?

Natural farming begins by sowing different seeds and determining what will grow on the land. What would the results be if this same method were used in an attempt to stop the deserts? That is, if the seeds of many different types of plants—as well as microbes, insects, and small animals, if possible—were sown and carefully watched. Anything that grew, no matter how small, would certainly provide some sort of a clue.

The first thing to grow in the desert might be a cactus, for example. If cactuses grow on arid land, then other succulents might grow there as well. Stonecrop, amaranthus, and wormwood might take at an early stage. Or perhaps plants of the lily family such as wild rocambole, leek, and garlic will be the first to grow.

On rocky hills and mountains that have been washed free of soil, plants of the grass family such as eulalia and cogon should take root in cracks and fissures in the rock. When these grasses begin to put down roots, vines such as Boston ivy and creepers spread out and cover the rock face. I have already confirmed that the stock varieties of such crucifers as shepherd's-purse and *daikon* grow quite well even in the desert. Striate lespedeza, Scotch broom, and certain other plants of the pea family also should thrive in sandy soil. Once things get to the point where *kudzu* vines begin climbing small shrubs and trees, the soil will become enriched and recovery to a green land will be rapid.

Before counting on glossy-leaved angiosperms such as the camphor tree, chinquapin, and oak, and gymnosperms such as cryptomeria, Japanese cypress, and pine, efforts must be devoted to increasing smaller organisms such as ferns, mosses, lichens, and soil microbes. Following this, plants resistant to drought such as the sesames and minor cereals will surely thrive, as will also gourd family vegetables and green manure plants.

Only God himself knows if this vision is just idle dreaming and whether good ideas for halting the encroachment of deserts can indeed arise from close observation of the sober and patient earth. In any case, we must first begin by trying. Changing the desert lands we already have into green and fertile plains will not be easy. Nor will it be a simple matter to halt the progressive desertification we are seeing around the world. Yet altering the vegetation and reforming agricultural methods is the basic approach to be taken. Circuitous as it may seem, this is, I believe, the quickest road open to us.

One thing that shocked me as I traveled through Europe was that even the rich, green landscapes which look so beautiful at first sight are artificial. With the strong conservationist sentiment that exists in Europe, I would have expected nature to be more fully protected. But the nature that has been preserved there is not true nature; the earth has in fact been ruined. Throughout my travels in Europe, I pointed out that this was a consequence of the errors in Western methods of cultivation. I believe that the day is approaching when natural farming will serve as one way of reviving the land.

Organic Farming and Ecology Are Self-Defeating —

If mistaken agricultural methods are responsible for the decay in European land, then unless those errors are rectified, both the rapidly declining nature and culture of Europe will be beyond help. It is generally thought that adequate measures are being taken to stem this decline, but is this in fact so?

In Japan as well, the issue of environmental conservation came to the fore in the early 1970s. With this, word of natural farming and organic gardening spread. But in spite of the high expectations made of it, organic gardening is not very different from scientific agriculture. In its present form, organic gardening is simply a return to animal-based farming and to the use of manure and compost. Because organic methods are essentially the same as those traditionally used in Japan, these can be of little help in restoring true nature. Not only that, if anything, such methods assist in the destruction of nature. True, organic farming does act as a brake, but since the brake is acting upon a broken wheel, this only compounds the danger.

If I may be quite frank about it, although organic farming appears to serve the cause of natural conservation, on reviewing developments over the past decade or so, this has not been the case. I began selling my mandarins directly to consumer cooperatives in Tokyo about ten years ago; maybe another ten years before that, some people I know got together and organized an organic farming association. Compared to back then when things were just getting started, it would seem as if the natural food and direct distribution movements have made some progress. But this has really caught on only among a small number of people. During the past decade, instead of moving toward the preservation of nature, the world and society at large has—just as I feared it would—continued on a course of relentless destruction. Nothing has been stopped. People living in Tokyo have not approached closer to a natural diet; if anything, their diet has become more unnatural—even anti-natural. In the space of these ten years, the assault on nature has proceeded at an accelerated pace, producing wanton destruction of the land and further debasement in the quality of man's diet. We can afford to wait no longer.

I think that the problem lies in people's willingness to believe that, with the clamor over natural diet, the development of organic gardening, and the slowing—however small—in scientific agriculture, things have been getting better. It is my belief that the arrogance—and failing—of scholars lies in their thinking: "If there is a right and a left, then a balance can be achieved and things worked out; as long as ecology exists and we have ecologists around, it will be possible to save nature." This is precisely what I mean when I say that halfway measures won't do.

I had an opportunity once to meet Professor Akira Miyawaki of Yokohama National University at a general meeting of the agricultural cooperative associations in Japan. Professor Miyawaki reported on pollution damage in cryptomerias along hiking trails at the base of Mount Fuji, stressing again and again that "nature must be protected."

After his talk, I spoke up: "Professor, if you think that the plant ecologists can protect the ecology of Japan's mountains and forests, you're sadly mistaken. It's not the plant ecologists who created the sacred groves of the local village shrines, you know."

The professor had a strange look on his face. I learned why later when I had a chance to read a book he had written; I found that he makes frequent mention of these shrine groves. I suppose that he expounds on ecology knowing full well the limits of plant ecologists and argues strongly for a revival of those groves. The problem is that the general public, taking false comfort in the thought that the professor and his colleagues will protect nature for them, becomes an unconcerned bystander in the destruction of nature.

Let me illustrate with an example. People who are told: "There's no one around to treat you if you get injured—there are no doctors on this island," are bound to take care of themselves. But try telling them: "We've got a surgeon and an internist on the island, so you can rest assured that you'll be well taken care of should anything happen." Do that, and people will cease to look after themselves. The more they hear talk of plant ecologists and conservation groups protecting nature and the environment, the less concerned people become about destroying nature. Now that we have environmental conservation groups in Japan and a national Environment Agency, it is as if people were saying, "Leave the fire up to the firemen and the arsonist up to the police." With its fixation on tourism and leisure, the public is calling for more high-speed roads and bridges. It looks as if the natural destruction of Japan will continue yet for some time to come.

Reviving the earth, halting the growth of deserts, and conserving the environment all can be achieved not by doing something, but by seizing an opportunity for restoring nature that requires nothing to be done.

The sacred groves of Japanese village shrines grew into natural woods only because, calling the rocks and trees there gods, someone attached sacred straw festoons to these, saving them from the ax. The ax and saw are the worst; their appearance marked the start of the destruction of nature.

Forty Days in Africa

Two summers ago I spent forty days in Africa, then last summer I made a second trip to America, where I toured the Pacific and East Coast states for close to fifty days. My reason for going was to sow seeds in the desert. I went because I have a great dream of revegetating the deserts and turning these into lands rich in food with natural farming.

It all began seven years ago when I boarded an airplane for the first time and was astounded to see the American deserts from the air. I started kicking up a row over this, vowing with some young people to work to revegetate California and even running some tests. News of this got to the United Nations, and I was invited over by an officer active in the Plan of Action to Combat Desertification. What really got me started was when he asked me if I wouldn't come up with some ideas for greening the deserts.

He was not joking. In fact, I felt almost embarrassed when I saw just how earnest he was. In the seven years that have elapsed since, the problem of desertification has never left my mind. When I stop to think about it, however, there really was nothing to it at all; the natural way of farming that I have practiced now for more than forty years could serve equally well as a means for stopping and revegetating the deserts. This will naturally transform a ravaged area into rich land where abundant grasses and trees grow in harmony with crops. And because no fertilizers or pesticides are used, I began to think that this might be useful also as a low-cost method of farming arid lands.

When NHK broadcast a program entitled "Sowing Seed in the Desert" on this idea of mine, some housewives in Tokyo—joking all the while at this quixotic vision of mine—

wasted no time in collecting funds to cover air fare, so I decided to fly to Africa.

My intention was to take with me 600 kg of *daikon* seeds donated by the Murata Nursery in nearby Matsuyama City, another several hundred kilograms of vegetable and grain seeds sent in by many individual contributors, and two hundred seedlings of Japanese fruit trees, and try seeing what will grow in the desert.

But when I attempted to carry out this plan, I learned that I had nowhere to go. Although aid to Africa in the form of food, clothing, and other articles was being offered from all quarters, strange as it may seem, not only did no organization exist in that vast continent for providing direct agricultural guidance to farmers and peasants, no one was apparently working over there in that capacity at all.

After searching all over Tokyo, however, I learned at last that several young volunteers from Japan were trying to set up a 75-acre farm for food assistance near Ethiopian refugee camps in a remote area of Somalia, so I decided to head over there.

But I had to wait a long time for a visa. Donating seeds and aiming for peasant self-sufficiency through natural farming went counter to the national policy of the Somalian government, which was trying to encourage the production of cash crops on large plantations. In the eyes of the government, what I was proposing amounted to rebellion. I received warnings and threats: "Giving seeds to a nomadic people and telling them to become farmers will be a blatant act of contempt. If the secret police find that you've taken even one photo, you won't be allowed to return to Japan. . . ." All this totally stumped me, yet, somehow or other, I was able to depart for Africa.

I took the plane at Shikoku and, after transferring at Tokyo and later points, eventually arrived in Mogadishu, the capital of Somalia. From there, I boarded a small Cessna and headed for the backcountry. With my fear of heights, I worried about arriving in one piece in such a tiny airplane, but when the plane took off, a splendid panorama of the vast savanna dotted here and there with thorny bushes opened up below us. The sight of the isolated round huts of nomads shared also by camels and goats made me feel as if I had wandered into some kind of fairy-tale world. When I reflected on my good fortune at being able to gaze upon such a wonderful vista at my age, I felt that I would have no regrets even should our little plane crash on the way. Later, upon returning to Japan, I learned that a Japan Airlines flight had crashed at about that time.

I was especially happy to note that, wherever I looked from the air, I could see stream and river beds. In general, I noticed also that there was at least one live spring within any field of view. What this meant was that, surprising as it seemed, there was a good deal of ground water even in the savanna.

It was with surprise that I later learned that two great rivers, each at least a thousand kilometers long, flow around the year through a desert that purportedly has less than 300 millimeters of annual precipitation. What is more, the desert starts right at the banks of these rivers.

But this desert sand contains clay. If there is clay, then there is hope. I told the young people of Somalia, "This earth is so young that it is sleeping. The land must be awakened from its slumber with natural farming."

In the region that I visited, hundreds of thousands of refugees from Ethiopia are living in temporary huts. These hardly qualify even as huts, appearing to be little more than large bird's nests made of thorny branches propped up against each other. The refugees poked fun at themselves by calling huts covered with dried leaves or a single, battered rag "luxury housing."

The food in the camps consisted entirely of wheat from the U.S., old rice from Japan, and macaroni from Italy. There were no vegetables. A typical hut had just one pan, one knife, and one cup as tableware. This was often the sum and total of a family's possessions.

But I did not come across the sort of pitiful scene so

common on television of weak, malnourished children lying down and covered with flies. On the contrary, I found cheerful, bright-eyed children running about and playing happily.

The refugees are clear about how they view food aid from abroad. "America's generosity is aimed at turning us into a bread-eating people. As for Italy, it figures that by giving us free macaroni for three years, we'll turn into macaroni eaters," they laugh.

In Tokyo, I saw a poster that said: "Listen to the voices of the African refugees wandering about in search of food and water." The photograph in the poster was taken of people returning from a cool, pleasant evening by the riverside. Even the single pieces of issued clothing they were wearing appeared to be beautiful garments fluttering colorfully in the wind. Or was this just my imagination? A professor who came to visit the farm for a short while during my stay in Somalia said that he had never seen such a hellish place anywhere before. To me, however, I felt as if I had learned something deep and vital: that it was such a place as this with nothing at all that was the entrance to paradise.

But I have to admit that the savanna landscape, bereft as it is of any growth but thorny acacia bushes and a few poison plants inedible even to goats, is a bleak and dreary place without colors or contrasts. There is no denying that I wished to see a bit of green vegetation here.

Several farms had already been built with foreign aid in this region, and attempts made to develop food sources locally. I was told that at the first of these, a farm built with Russian aid, long irrigation canals on great levees had been constructed but the farm had since been abandoned and these canals left to crumble. I learned also that U.S. aid had been used to build a large, 500-hectare (about 1,250 acres) farm, but that all the low bushes and shrubs had perished before the bulldozers and attempts to plant crops had been unsuccessful. When we visited the place, it looked like the sad, empty remains of an abandoned airfield.

The only exception was a French farm where tomatoes and onions had created green fields. But this was located on a riverbank with large trees, and those actually growing the produce were Somalian farmers. One farmer holding a bunch of onions in his hand approached me with a look of concern on his face. Upon examining the plants, I understood why: they were swarming with mites.

I met a couple of young people connected with civil engineering working at the Japan International Volunteer Center (JVC) farm. They said that over the past two years, they had experienced failures when the oil needed for running the irrigation pumps ran out. From what they told me, prospects for that year seemed little better.

Generally speaking, when a large farm is built in the savanna, the land is first leveled with bulldozers. Next, high earthen levees are built out from the river at a perpendicular angle and irrigation canals are dug in the top of these levees. Crops are grown by pumping up river water and pouring it sequentially onto fields divided into small sections each about one are (a hundredth of a hectare) in size. Because this results in the same build-up of salt as when water is sprayed onto a salt field, I doubt that such a method can be sustained for very long. I thought the Japanese way of building irrigation canals to be the best, but I found it hard to understand why no use was made of water wheels and windmills.

If I were to set up a natural farm in the desert, I would apply a method that might be referred to as "plant irrigation." Instead of making canals, what I would do is to create a green belt consisting of various types of trees, using the roots of these trees to draw in river water underground and gradually penetrate dry ground away from the riverside. Or I might employ special plants to draw up and store this ground water so as to make it available for use. While in Somalia, I was able to gather samples of several plants that could be of help in combating desertification and removing salt from the topsoil.

True, natural farms have employed a particular sort of layout that features the interplanting of fruit trees with other trees and takes into account rotational relationships such as the mixed planting of green manure, vegetables, and grains beneath the fruit trees. However, my ultimate desire was to run tests on methods of scattering drought-hardy coated seeds over arid lands by airplane. Unfortunately, this was not understood in Somalia. I learned only that without the co-operation of the government, one can do nothing.

I decided to visit nearby farmers and refugee camps, hand over seed and try having people see whether they would create home gardens. The children, drawn by curiosity, immediately thrust out their hands. When I gave them seed, they shrugged their shoulders, saying, "Wangaranai" (I don't understand). Gesticulating, I showed them how to make a furrow with a stick on a sand dune, drop seeds in the furrow, and apply river water to the seeds for three days. Two or three days later, a group of about twenty children came over. Saying, "Kai, kai" (Come, come), they beckoned me to follow them. I went with them down to the river's edge where they stopped at a small, round garden a couple of yards in diameter. There I saw seedlings of *daikon*, onions, cucumbers, and beans sprouting up all over the place. I became all excited because I hadn't expected such an easy success.

From that point on, older youths at the camp began coming for seed. Several days later, no more than five minutes after I had arrived in one village, tens and hundreds of women and old people gathered around me. It was like selling hotcakes. Before I knew it, the two large bags of seed in the car had disappeared.

Together with children and others, I even scattered seeds in clay pellets over a rock-strewn stretch of savanna. Then, with forty or fifty people I mixed rice seed with barnyard and proso millet and scattered these, running water over the land and having everyone stamp down on the scattered seed. The purpose was to hide the seeds in the sand, but when I told them that we were planting the seed, they were tickled by this and had a ball doing it.

So I was unable to carry out my original plan and instead just mixed and planted a small portion of the seed I had brought with me. Practically all of the vegetables germinated easily. As for the fruit tree saplings, I learned that these would take root, but headed home without being able to see for myself the results of full tests.

Following my return home from Africa, I learned that large *daikon* were grown, and that little vegetable gardens have sprung up around many of the huts. Little by little, I am told, green vegetation is appearing in that refugee camp that was so bleak and desolate when I saw it. This vegetation is most abundant at the river's edge and in the areas where we planted seed.

Of the fruit trees, reports have it that the lemons and other citrus trees, and the persimmon and pomegranate trees have grown especially fast, some attaining to a height of about five feet in just a half-year.

As for the bamboo I planted by the side of the river in the hope that this would provide clues to revegetating the deserts, the *danchiku* and rushes have already grown to about 6 feet or so. Acacias and silk trees grow rapidly, so the roots of the saplings we planted probably have already reached the ground water. The banana and papaya saplings planted below these should also be growing very well.

While it is vexing not to have all the details, in spring 1986 the Somalian government recognized the benefits of home gardens. As a result, the young Japanese man who showed me around during my stay there was singled out and promoted as head of JVC operations in Somalia. A proposal he made was accepted at the United Nations, which has started up a home garden project in Somalia and Ethiopia with over a half-million dollars in funds. Judging from this, I would say that my trip served a useful purpose.

What makes me especially happy is to hear that those

refugees who had no green vegetables to eat—just wheat flour and macaroni—have begun eating *daikon* leaves and even the white roots that they refused to touch at first. If everything goes well, in two or three years they may be enjoying cool breezes in the shade of the fast-growing palm and banana trees.

Still, Africa is just so large that even if green vegetation and forests are restored to one part of it, this really accomplishes very little. Yet when I rack my brains searching for some way to carry on large-scale tests for total revegetation of the desert, I have to remember that Somalia, after all, doesn't even have a postal system. This is a country that only developed a system of writing about twelve years ago, so written communications are virtually impossible. One young woman from Japan did at last go this past spring, both as my contact and to sow seed, but word has it that this too has not turned out as I had hoped.

Why the Tragedy in Africa? ━━━━━━━━━━━━━━━━━

I have only briefly glimpsed one small corner of Africa, so I am not really in a position to say anything, but what I think is that people in developed countries should all get out of Africa—that is, other than those providing some temporary aid. The reason is that true happiness does not lie in the direction in which the advanced nations are headed.

Not only do Africa's people have an excellent capacity for independence and self-reliance, they know what the ideal society of the future is.

From the little that I have seen, I can say this about the current African tragedy:

1) The internal cause is as follows. Nomadic people were not content with hunting wild beasts and birds, and so with the development of the notion of accumulating

wealth, they began to raise large numbers of goats and cattle. I suspect that this is when vegetation began to rapidly disappear. In other words, it is quite likely that the desertification of the mountain forests and plains in Africa began when the grazing of animals by nomads reduced the variety of animals and plants living on the land, upsetting the balance of nature.

2) External causes include the system of national boundaries introduced by outsiders and the natural wildlife preserves forcibly established under the pressure of the conservation movement. These destroyed rules observed since antiquity (some of which can be found in the Koran), and brought an end to the free movement of nomadic people. The resulting loss in the time between grazings essential for natural recovery of the land can be regarded as an external cause that threw nature into chaos and triggered desertification.

3) One calamity has been the abandonment by farming peoples, in the name of agricultural modernization, of time-honored methods of self-supporting agriculture that enable them to produce many different types of grains and vegetables. Such native farms have been replaced with large plantations growing cash crops such as coffee, cocoa, sugar cane, and cotton, a change that caused rapid desertification of the flat plains.

In short, without a plan and policy to revegetate the African deserts, the people living there cannot become independent and self-sufficient. Without the ability of a people to be self-supporting, an ideal society cannot be built.

Before becoming contaminated with the materialistic thinking of the developed nations, the ancient "Great Spirit" must be revived and a spiritual culture developed. The first step in attaining this is the establishment of domestic life, and the quickest and most effective means of doing so is through natural farming. As a life of autonomy and self-sufficiency is built up through natural farming, revegetation of the deserts will surely follow.

However, such means will be applicable only close at hand by individuals. It will not do to attempt first to revegetate small parts of the great, wide desert, then gradually expand this effort to other areas. Not only is there no time for this, such an approach would stand little chance of success. The only way in which an entire vast region can be revegetated at once is to scatter the seeds of grasses, trees, and crops suitable for greening the deserts from airplanes. We still have a long way to go.

America Revisited

As I was wondering if there wasn't something that could be done, I received a very polite letter from a high-ranking tribal Indian chief in America inviting me to come establish a natural farm on an Indian reservation in the American grasslands. I answered that I was interested, and later received formal invitations from two West Coast colleges asking me to attend international conferences on natural farming as a guest speaker. The letters said that if I were to come, a schedule of about two months would be drawn up for me with people active in natural farming serving as my guides and interpreters.

I decided to go, thinking that the trip would serve a purpose if it were even a little helpful in resolving the problems that I had been unable to make any headway on in Africa. I departed from Narita alone, heading for Seattle on a Japan Airlines flight. From the moment I arrived at the airport in Seattle, everything was taken care of for me. I simply did as I was told by those around me. They had done a thorough job of investigation beforehand and took me around to any places they thought would be instructive to me. Moreover, they very skillfully set up a tour schedule that allowed me to say everything that I had to say during my trip.

According to the itinerary that had been drawn up for me, I was to travel south from the state of Washington, tour Oregon and pass down through California before flying off to New York. The focal point of the tour would be international conferences at two colleges and lectures at various schools and organizations along the way. Also included on the daily agenda were one or two visits to farms and practical demonstrations of natural farming techniques, so each day I spent at least 2–3 hours, and at times up to 5–6 hours, traveling around by car. I even boarded light aircraft on four or five occasions. Since I was also interviewed by newspapers, radio, and television during the course of my stay, this was quite a hard schedule that left me hardly an hour free to myself. I was more than a little amazed later on to realize that I had been able to keep up this pace for close to fifty days. I am easygoing by nature, so each morning I would simply ask where we are going today. Even when it came to lectures, I was often informed of the type of audience I would be addressing and how long I had to speak only shortly before the scheduled time. As they say, "Ignorance is bliss." Such an outlook certainly made things easier on me. I would get up on the podium and, after looking out at the faces of the people in the audience, speak about whatever crossed my mind, so no matter how long I went on talking I never got tired. Because it was a fascinating and enjoyable trip, the fifty days passed in a flash.

American Agriculture Seven Years Later ——————

My impressions after this last trip around America are totally different from those I had seven years ago. The climate surrounding agriculture has changed completely. In 1979, I came thinking that America was a green land. When I found instead parched, barren, land, I sounded off about how nature in America was artificial and how farm crops there were petroleum-based products. I warned also that the future of American agriculture was in danger. My second trip in 1986 showed me that I had been right on the mark.

About 30 percent of North America consists of arid land; the central grain belt, which has been severely depleted, accounts for another 30 percent; and another 30 percent is made up of green plains. Forests with trees large enough to harvest for lumber cover only a few areas such as the state of Washington, and constitute perhaps about ten percent of the total land area.

In a word, with more than half of the country already a desert or fast approaching a desert-like state, the situation seemed to me even more pressing than in Africa. Yet, most Americans have very little awareness of the extent of this change over their national lands. They find it normal, for example, when no rain falls during the summer in California and the grasses dry out, turning the region into a bleak, yellow land. They even call this a continental climate. They were surprised when I told them that those grasses are dormant, not dead, and that while the summer in Japan is also hot, this is the season of deepest green.

Falsely reassured by the expanse of their country, many Americans show no desire even to protect the land. I've heard that a third of America's farmers are quitting farming, and that this has triggered a rash of bankruptcies by local banks. Everyone seems convinced that this tragedy in America's farmlands is caused by a sluggish export market for farm products, but there is little reason to believe that farmers will ever be well off when their land is in decline, no matter how much they produce crops with "petroleum rain." Sad to say, the general public does not seem to have taken this to heart.

Agricultural scientists and entrepreneurs in agribusiness all make light of the situation, claiming that even in an arid land that has lost its natural energy, they can grow crops any-where and anytime as long as oil energy and water is sup-plied. Japanese scientists and farmers concur, believing that this is what modern agriculture is all about. They have begun to treat the soil as if it were an inconvenience, but I wonder if they haven't forgotten something very important.

Today, the pivot farm is a fitting symbol of American agriculture. A mammoth sprinkler system a half-mile long rotates about a circular field continuously sprinkling water. Only the interior of the circle is deep green; because these fields are located in the middle of arid land, they are easy to see no matter how high one is flying. But since water for the sprinklers is drawn up from depths of hundreds of feet, it should come as no surprise that this practice results in salt accumulation on the soil surface. Such agriculture amounts to the "throwaway" use of land. After 5–6 years, the sprinkler system is moved to adjoining land and the abandoned field becomes a true desert.

I believe the desertification of American land to have been triggered by basic flaws in farming methods. But compounding this problem is the fact that on abandoned land which has become a dry grass prairie, the soil temperature rises greatly on account of reflected heat. I did some investigating of my own and found that the soil temperature on exposed land covered with dry grasses is normally 20–30 degrees Celsius higher than in areas covered with green grasses or near woods. When a spot adjacent to lush vegetation is 30 degrees Celsius, the temperature on a stretch of yellow grasses rises to 70–80 degrees Celsius. The state forestry bureau researchers who ran the tests were themselves astonished at these figures. I had suspected that the reflected heat from arid land affected surrounding areas, but this made me realize once again that we must waste no time in revegetating the deserts by sowing *daikon* and other hardy seeds.

In any case, I think there can be little doubt that the poverty of American farmers who make less money on 500–700 acres than do Japan's farmers on 3–5 acres is basically a consequence of the decline in the land resulting both from their abandonment of nature and dependence upon artificial oil energy rather than natural energy, and the appropriation by merchants of both the means of production and sales.

American agriculture must be revived and a program implemented for combating desertification. This revegetation of the deserts can become the starting point for a revival of agriculture. I was told that an American agricultural reform was in the offing on the Pacific coast. As I traveled through the Western states, I myself sensed the first signs of a new agricultural revolution emerging among the farmers of this region.

The Outdoor Food Markets ────────────────

These farmers have relinquished the general tendency in modern farming toward very large farms and begun to move in the direction of Eastern thought and natural farming. Nor has this been simply a means for shaking themselves out of a depression.

At all of the lectures I gave, I received a big welcome as the author of *The One-Straw Revolution*. However, it seemed to me as if their warm reception bespoke their expectations of me as a new Eastern thinker rather than as a teacher and leader of natural farming. Today in America, a revolution in spiritual awareness appears to be underway not only in agriculture, but in all other areas as well. I sensed the same current both in the cities and farming communities. It seems as if every city and town in America now has a store or small market that specializes only in natural foods. Some of these shops are run by agricultural cooperatives and are doing brisk business right next door to the supermarkets. This could well be the start of a new distribution system.

What caught my attention in particular were the Sunday and morning markets in the cities. Generally situated on a garden-like plaza such as in a park and decorated with colorful banners, these included exhibits by street artists and stalls selling souvenirs and toys. Lively music performed out in the open added to the atmosphere. Unlike at Japanese markets, the stalls were nicely decorated and pretty salesgirls cheerily hawked their wares to passers-by.

There were many types of food stands, offering Italian cooking, Indian cooking, and French cooking. Westerners wearing Japanese *happi* coats were busily rolling *sushi* with their hands and preparing various *tofu* dishes (many not found in Japan and quite delicious) on small plates. What was unique about these dishes was that they were all improvised and prepared by amateurs.

At the center of the market were stalls heaped high with

fruit, vegetables, fowl, live fish, natural breads and cakes. What was special about these markets was that the products sold there were limited to natural foods. It is generally decided, with the consent of the shoppers, that standard supermarket products be kept out of the market.

The Sunday markets being the exclusive domain of natural farmers and amateur merchants, it was no surprise that the salesgirls were so full of pep and enthusiasm and that the diverse goods, many of them original products thought up by amateurs, sold freely in such an atmosphere of amusing and delightful animation. This is a place of creation. Without question, greater variety is to be found in the products here than in the supermarkets. And because these are fresh, the people of the city enjoy shopping here on Sundays. They find it pleasant to eat out, lie down a bit on the grass and, after buying a week's supply of fruits and vegetables, drive on back home. I can understand why these markets are becoming so popular.

As I was wandering leisurely through this marketplace drinking in the sights and sounds, from time to time someone

There never was any East or West, high or low.

in a stand I was passing would call out, "Hey, you're Fuku-oka, aren't you?" Regarding my *monpe* garb with curiosity, they would give me what seemed to be the most unusual apples and eggplants. At a few of these markets—such as those in Davis, California and Eugene, Oregon—it seemed as if practically a third of the people knew me. They kept handing over souvenirs to those traveling with me; before long, I too was drawing my crude "philosophical cartoons" on cardboard signs with a felt-tip marker and passing these out. People were delighted with the sketches and would immediately put them up as signs for their stalls.

These cartoons of mine were often used as posters at the entrance to university lecture halls and in front of natural food stores. To my astonishment, I later even found them being sold on T-shirts. Now why should these drawings be so popular among Americans?

The hearthside is the universe ... and the universe is a dream within a pot. (Above the pot is the character 空 representing emptiness.)

The sight of these open-air markets is clearly Oriental. It seems to me that, having tasted the open ambience of the marketplace, the Americans, who have long prided themselves on their individualism, have begun to associate with each other in a more frank and magnanimous way that is typically Oriental. This also seems suggestive of the future inclinations of America's city-dwellers.

The Growing Popularity of Japanese Cuisine ─────────

One other phenomenon that must not be overlooked is the natural foods boom that is sweeping the entire country. At the roots of this lies Eastern thought. I was aware that the long-standing efforts of Michio Kushi in Boston and Herman Aihara in California—both of whom invited me over during my first trip to America in 1979— to popularize natural foods had helped trigger a boom in Japanese cooking. But on my last visit to America, I was able to see for myself that Japanese cuisine has firmly established itself in America as a type of ethnic cooking that is both delicious and good for the health. Today, in some areas, it is second only in popularity to Chinese cooking.

Because the raw materials are good, the taste of food in the big-city *sushi* and *tempura* shops may well be better than what can be had in Japan itself. Even if you look at the way Americans eat Japanese food, they seem very much in their element, from the way they select the ingredients for *sushi* to their method of rolling the *sushi*, the materials they use for *tempura*, and how they make *tofu*. Why, I've even heard Americans comment on whether a buckwheat noodle broth is good or not. I thought Americans were terrible cooks and undiscriminating about taste, but today things have gotten to the point where, even in ordinary kitchens, people are cooking and serving rice-based meals—even dishes such as brown rice gruel, modified and prepared in American style, of course.

When things reach this point, then the farming population

must answer to the needs of consumers and ensure the supply of the desired materials. With the improvement of culinary skills and the demand for abundant raw materials, it is to be expected that demand for food produced by natural farming will begin to emerge as an outgrowth of the popularization of natural foods.

Up until recently, the lack in Europe and America of the same, broad selection of soft, leafy vegetables grown in Japan, the scant number of root vegetables in the diet, the generally restricted range of raw materials, and the very limited culinary skills all combined to give Westerners an underdeveloped sense of taste. The same varieties of California tomatoes and oranges are sold at the same time throughout the entire United States. The American women and men, parents and children who can without the least concern eat the same food day after day, and the American farmers who, believing that producing nothing but soybeans in the soybean belt and nothing but wheat in the wheat belt with large machinery on enormous farms is the height of rationality, have stoutly defended monoculture are one and the same. Much as this might be shrugged off as a continental temperament, it cannot be denied that many have begun turning to the delicate and subtle Oriental flavors, farming methods, thought, and way of life.

The driving force that unleashes a major reformation of American agriculture could very well be such changes in the diet of ordinary people followed by the spread of home gardens. And this just might lead to the growth and development of natural farms.

Natural Home Gardens

Scholars from all over the world gathered for three days at a hot springs resort called Breitenbush in Oregon to take part in a colloquium on nature. It was an enjoyable occasion, part of the time being spent soaking in open-air baths and saunas heated with geothermal water. This also served as a preview of the upcoming international conference.

The resort was situated in a large virgin forest. The main buildings, such as the auditorium and dining hall, were scattered here and there, and there were also dozens of bungalows for lodgings. In addition, tucked off in one corner of the grounds was a natural garden for growing the produce used in the dining hall kitchen. I was told that after the person in charge of the garden had gone, this garden had been left alone.When I heard this, I decided to hold a workshop at the garden before giving my lecture. The first one to sow seed here had been a young Japanese man named Katsu. It so happened that he was traveling with me as an interpreter so he had a chance to take another look at the garden that he had started working on two years earlier.

Katsu told me that, making use of natural farming techniques, he had sown clover over the entire surface, then scattered many different types of vegetables in little plots within the garden, but that after he had left the area, the garden had grown wild like this and failed. It was true that, at first sight, the place was overrun with weeds and looked in total disarray. Anyone would take this for a failure. But when I entered the garden to take a closer look, I found that the clover had spread throughout the entire garden and the number of weeds was not as large as it had appeared from a distance. Indeed, all kinds of vegetables were growing to very respectable sizes.

I summed up what I saw to the workshop participants as follows:

"This is just fine as the first step for a home garden prepared through natural farming. Man sowed the seed the first year, nature made some adjustments the second, and in the third year God built us a natural garden. The proof of that is clearly evident in the process through which this garden has gone. Maybe I shouldn't put it this way, but because you left this garden untended, Katsu, the many vegetables bloomed and the fruit set. This attracted birds and mice, who ate the seeds and spread them throughout the garden; they essentially sowed the seeds of the different vegetables you planted everywhere. As you can see, this garden has a range of growing conditions; there are dry spots and wet spots, areas of poor soil and shady spots. As a result, not all the seeds that fell germinated. You might call it the 'survival of the fittest.' The following year, only those seeds that fell at the right time and in the right place germinated. There is no getting around this seemingly disordered state in which those which are doomed to perish do so, and only the survivors thrive. But nature eventually shows us in this way where each plant will grow best. All one has to do is look at this and follow nature's example. Because clover has spread over the entire field, all you need to do is to suppress a few 'problem' weeds and an outstanding garden will develop naturally. There just happens to be a clump of wheat and a clump of naked barley that were not sown by man here. The heads on these have more kernels than normal. Judging from this, I would imagine that mixing in naked barley and other grains with the seed that you sow here might produce some interesting results. Garlic, for example, is growing very well here so you ought to get good results by growing naked barley in the winter and spring, following this with plants of the gourd family, then planting, say, onions and broad beans where you had cucumbers and melons.

"In any case, although this field may appear to be a dismal failure, I would say that, thanks to a stroke of good fortune, it is doing very well for natural farming."

When I explained things in this way, the members of the workshop were delighted, saying that this gave them a good idea of the methods and theory behind natural farming. But best of all, this restored Katsu's self-confidence.

I had felt until then that discussions of natural farming tend to get abstract and vague, making this difficult to understand well. But when I went to America and gave some practical instruction in my methods, I found that people there catch on very quickly. Afterwards, when I traveled to many different farms and actually demonstrated to people what I meant, even those who remained unconvinced by the theory were ready at once to give the method a try. The reason for this occurred to me later: in Japan, even those who do tend a home garden generally do so either as a pastime or to produce food for domestic consumption. At most, they desire only to be able to eat food free of pesticides. It seems to me that, in contrast to such meanness of spirit, Westerners get into natural farming because of an intense desire to become one with nature, to plunge into the heart of nature and see what one is capable of. Because they bring with them the tolerance born of the desire to play together with nature, they are not put off by the sight of a few weeds. So what to the Japanese might be an unattractive, poorly tended garden all gone to the weeds would to many Westerners be a picture of natural harmony at its best.

Although that field may have many weeds, because there are also large *daikon* and enormous hanging eggplants, this is where the applause is due. The fact of the matter is that natural farming appears to have made a solid start in the gardens and among the small farmers of Europe and America.

A Natural Garden in the Backyard ───────────────

Unlike in Japan, yards in America are quite large. Many people with a mind to do it could grow enough produce in their backyard to meet just about all the food needs of the family.

I have proposed that people in Japan grow their own food and that a family have a quarter-acre of land on which to live, but in America I learned to my surprise that, although zoning laws differ from locality to locality, in many areas one cannot build a house on such a small piece of land. When the yard or land about the house is large, the living environment changes completely. If the Japanese, like the Europeans and Americans, were to get the urge to move to the mountains and enjoy a life of freedom and autonomy, Japan has quite a lot of mountains and it is almost certain that the bottom would fall out of land prices in the cities.

In California, high-speed roads cut across the dry, desolate prairies, but the moment one enters a city, large trees rise up everywhere so that one feels as if one were entering a forest instead. The roadside trees are all natural trees and grow so densely that one gets only a glimpse of the buildings through the line of trees as one drives past. The yards where one is barely able to see the house from the road belong to middle-class people. The wealthy live way back in the richly forested hills. There is no vegetation in the inner-city areas where the poor live. Most cities are tranquil, clean places full of vegetation, but upon leaving the city, one returns to the open prairie where there is no green vegetation. The opposite is true back home. In America, the most luxurious dwellings are located in deep or virgin woods. Unfortunately, the Japanese do not have the same spirit of independence as Westerners. The average American home has a spacious yard, natural woods, and a lawn, so when people set up vegetable gardens there, something more ideal than is possible in Japan may arise.

The ideal natural farm that I envision has vegetables and grains growing beneath a mix of fruit trees and other trees. It is a farm where river and mountain, trees and grasses exist as one whole and harmony is maintained. When I say such things in Japan, people look at me as if I had lost touch with reality, but in America I find that putting things in this way makes it easier for people to get a solid understanding of what I mean and apply these ideas in practice. For example, when I said that various types of fruit trees ought to be inter-planted with roadside trees in the cities and that vegetables should be planted instead of lawns and flowerbeds so that people walking by can pluck and eat roadside fruit or pull up a couple of *daikon* and take them home, the idea aroused people's interest wherever I went. This would create "home garden" boulevards everyone could use.

As for lawns, when I talk about how, when clover and *daikon* seeds are sown over a lawn, the clover will win out over the grass in 2–3 years, enabling the *daikon* to grow within green manure, those who get the biggest kick out of this and go right ahead and try it are Japanese wives, Chinese, and other Orientals living in America. Although they agree with my reasoning, Americans, on the other hand, laugh when I bring this up because they know, from the close attachment of people to the serene and dignified "lawn culture," just how great a revolution this would be. As long as this problem remains unresolved, the potential of natural home gardens in America will be limited.

What Is This "Lawn Culture"?

The goal of the average American in life appears to be to build up some savings, live in a large house in the country surrounded by tall trees, and keep a well-manicured lawn. An even greater source of pride is to own several horses.

I was told that lawns were a cultural outgrowth of the pastures on which aristocrats grazed their horses. In short, the ideal of people in the West is an aristocratic life. Living as they do in country woods, their lifestyle resembles that of an Oriental hermit. In fact, however, it is just the opposite: an extension of aristocratic tastes and nothing more.

If lawns are indeed a vestige of aristocratic culture, talk of rejecting the lawn is tantamount to asking Westerners to renounce pride in their past. This, of course, is out of the question. Yet, although lawns may be safe and pleasant for man, they certainly do not benefit nature. If nature is sacrificed to create these stretches of artificial green, then the "lawn culture" of the West is all a big mistake.

On this last trip of mine, I often brought up my argument

for rejecting the culture of the lawn, expecting full well to earn the hatred of Americans in the process. My reason was that, short of doing this, it is not possible to successfully promote the idea of home gardens.

If Americans really wanted to bid their cherished "lawn culture" good-bye and create natural home gardens, their home life would benefit greatly. When I spoke one night at a church on the University of California campus at Berkeley, some of the people there decided to give this a try in their own neighborhoods. I suspect that things may get underway first in the corners of the city where Japanese wives live and in Chinese neighborhoods because the Chinese enjoy making vegetable gardens. The trees and grasses know no national boundaries. When streets and neighborhoods arise without boundaries between the houses and yards; when trees on these streets bear fruit everywhere and *daikon* plants bloom wildly, this will also become an incentive to create natural gardens and farms. And who knows, this could even lead to the sowing of seed in the desert.

Lundberg's Natural Rice ——————————————

The campaign to restore nature to America has begun not only among small farmers but also on large farms. I am referring to the Lundberg natural rice grown in Chico, California and known widely throughout America. I visited the 7,500-acre farm owned by the Lundberg brothers seven years earlier during my first visit to the States. Harlan Lundberg had listened to what I had to say, then jumped up in delight, exclaiming, "Why, that's incredible! It's a revolution!" I had heard that a short while later he fired six tractor operators and began a natural farming operation but I had had no further word on how things had worked out since then. When I saw Lundberg again last year, he said, "Since meeting you, I've grown a lot more tolerant about things. Come and take a look. Now all four Lundberg brothers are using natural farming." Sure enough, when I went over to the farm, I saw four grain elevators standing there side by side.

"Three hundred farmers have gathered here today. We're going to take this opportunity to form a grower's association so as to further increase the amount of natural rice grown and sell it all over America."

Harlan presented me with a large gold medallion, saying, "This is a medal our company give to workers of special distinction."

After this, everyone drove in a long line of cars down a road passing through the Lundberg's extensive fields (all you see around you on all sides are rice fields). The cavalcade came to a stop at a cool, slightly wooded spot where we had an outdoor party and the farmers listened to me speak.

What impressed me was that, although I saw lots of millet growing in these vast fields thousands of acres in size, Harlan did not seem bothered by this. Nor did any of the many farmers who had come to take a look at the farm show the least sign of concern.

Had this been Japan, these clearly would have been failed fields; farmers in Japan would have said that if there's going to be this much millet growing in the paddies, they're not even interested. But the Lundbergs and all the others before me were totally unconcerned. When I saw this, I felt as if I understood what he had meant when he said that he had grown more tolerant.

While I marveled at how he had been able to look with equanimity upon this millet for six full years, I noticed that this very vitality of the soil was the reason for the success. Even with all this millet, the land was producing 16–18 bushels per quarter-acre (average for Japan). And since harvesting with large machinery poses no obstacle whatsoever, I suppose Harlan's calm and composure in the face of all this millet was not surprising.

Seven years earlier, he had been working to set up an organic farming operation and was growing rice only once

every three years (the land was left fallow a year and summer wheat grown the next). But after we met and talked, he
became able to grow rice continuously every year by means
of natural farming methods. Moreover, since the product is
natural rice, he can sell it at twice the price of ordinary rice
in America. It is understandable then that he has done very
well.

Moreover, he has overlooked nothing. He has interbred
many combinations of long-grained yellow and black stock
varieties of rice and sells these as delicious natural rices. He
announced plans to incorporate my rices as well, along with
other varieties, and, by working in cooperation with the 300
farmers, to expand activities throughout the U.S. And he
talked of how he had resolved in this way never to knuckle
under to big oil and the capitalists behind it, come what may.

The first time I went to America, my wife came along with
me and tried to restrain me, saying, "Do you think you
should be lending a hand to rice production in America at a
time when we have a rice surplus in Japan and the government is talking of cutting back acreage for rice?" Indeed,
after returning home from this second trip to America, I
found everyone in an uproar over the question of importing
rice from America. But the way I see it, rather than addressing the basic issues, it is showtime and both countries are just
trading clever jabs before the audience.

When I saw the enthusiasm of a farmer in Chico—at the
heart of California rice country—who said he wanted to
grow my new varieties of rice, I realized beyond a shadow of
a doubt that this would go a long way toward converting
Americans to a diet of natural rice. Before fretting about
pressures from abroad, there are certain things that Japan
also must do.

Sowing Seed in the Desert

My ultimate dream is to sow seeds in the desert. To revegetate the deserts is to sow seed in people's hearts. It means to turn the earth into a peaceful, green paradise.

Nature in our modern world is in rapid decline and confusion reigns in people's hearts and minds. We must consider this day as the Genesis. I suspect that the time has come to remake the world. If we were to gather up the seeds of the world's plants, mix them together, and scatter them all at once from the sky so as to turn the world into a green paradise where anyone can obtain food freely anywhere, the many problems borne by man would become resolved in and of themselves. To the grasses and trees, neither national boundaries nor human races exist.

In the summer of 1985, I traveled to Africa, where I ran some tests in which I scattered seed in the desert to determine how revegetation can be achieved. Arguing that it is possible or impossible will get us nowhere. Revegetating a desert is totally out of the question unless one sows seed simultaneously over the entire desert by airplane. Also, desertification is proceeding at such a pace that any other approach will be too little too late. Such a fantastic vision doesn't even rate a laugh in the island nation of Japan, but I feel that this is more than just a dream when I speak of it in America. There is every indication that this can be done provided the determination is there to go ahead and do it.

To give one example, as we were riding south from Oregon to California on a highway, we talked in the car of the idea of scattering *daikon* and green manure seeds directly from the car onto the parched wasteland along the road. One of the young people riding in the car took out a bag filled with

different seeds and said he was giving it to me. When I asked
who he was, I found that he was a taxonomist who was
collecting rhizobia and especially nitrogen-fixing plants. He
had heard me speak a few days before and had joined the
group traveling with me because he had been moved by the
idea of a campaign to sow seeds in the desert and wanted to
help. When we reached the beautiful high pass just before
entering California, he had the car stopped, handed me the
seeds he had prepared, and said, "Try scattering them over
here." I scattered the seeds from the top of the pass toward
the valley bottom. Everyone else riding with me began
doing the same thing, shouting out, "This is it! This is it!"
The seeds caught the wind and flew far off.

He was not the only one. In the states of Washington and
California as well, there are groups and individual botanists
doing dedicated research. This research is low-profile and of
no use to society. They collect the seeds of primitive vegetables and plants native to a local area even though these
have no cultivation value, and promised me that whenever I
give the word they would send me seed. Among others, the
head of the Paleobotanical Gardens in San Francisco, who
is familiar with Africa, also promised that he would collect
the seeds of plants suited for growth in the desert. I would
like to give one example here which supports my expectations
that Americans will indeed act.

In Ashland, Oregon, I gave a lecture one day at the university, and the following day visited and gave talks at two
farms in the area. When I had finished talking, one young
man stood up and said, "Are all of you just here to listen to
what Fukuoka's saying? I'm a pilot and I've got a plane for
scattering seed. It's even equipped with a gun for blasting the
seeds into the soil. How about using it?" When he had said
this, several dozen people near him began pledging their
cooperation in gathering seed.

Then one old woman stood up. "I've got a 750-acre piece
of wasteland that I'm going to let you have so give it a try

and see." It took all of ten minutes for the local people to decide to revegetate this entire region.

The south side of the valley here has turned into a savanna-like grassland barren of all green vegetation, but the opposite side of the valley has forested mountains and is a magnificent place famous for its scenic beauty. Just the thought of such a valley becoming filled with blooming *daikon*, is enough to set one's heart beating wildly.

I have something even better to announce. I had been hoping to hold talks with people involved in the Environment Programme at the United Nations, but had no luck with this because everyone was out on account of the summer vacation. However, when I returned to Tokyo, I was contacted by a Frenchman named Henri Lucy living in the African Congo who said that he had learned of me through the Kushis in Boston and wished to meet me. As a result of our talk, he said that he would propose in several countries that seeds be sown by airplane and would try to have the U.N. do this. It appears that he is the person in charge of such matters at the U.N.

The Kushis were told by Lucy's wife that her younger brother is an industrialist who owns a number of airplanes, and that he could provide about a half-year of support in case things don't work out at the U.N.

In any case, once things come this far, sooner or later someone is going to sow seeds in the desert. It looks as if my going around and saying that Reagan ought to use those bombers and space shuttles to rain down seeds all over the world rather than to fire missiles was not totally in vain after all.

Making the Change from Organic to Natural Farming —

With the continuing input of corporate capital, American agriculture will probably go on growing even larger in scale.

At the same time, advances from organic gardening to natural farming are likely to continue to be made by people with a proclivity toward natural methods. The problem, however, is that most people do not yet understand the distinction between organic gardening and natural farming. Both scientific agriculture and organic farming are basically scientific in their approach. The boundary between the two is not clear.

Inasmuch as the primary goal of the international conventions I attended on the West Coast was to arrive at an understanding of the current world situation and to ponder what direction future actions should take, participants examined specifically how various farming methods now in use, such as permaculture, organic farming, and various other agricultural methods founded on new concepts, relate to each other and how these can be made to act in harmony with each other.

The way I see it, and perhaps I am biased, the only way is to follow the road to nature as perceived from afar. In so doing, techniques that surpass mere technology will be established. That is my view.

Although there are still many different forms and names for it, it is clear that my "green philosophy" serves as a foundation.

It is certainly fine to gradually move from organic farming toward an anti-scientific way of farming, to aim for a sustainable and permanent method of farming, or to attempt to return to nature while enjoying life on a designed farm. However, this must be more than a narrow technique; nor should it be adopted merely as a passing fad. The thinking and outlook of natural philosophy must be at the core of any successful effort to establish a form of farming that would become a truly permanent Great Way of agriculture.

Although I stressed this at the international symposium, I was able to ascertain whether such thinking actually stands a chance of being accepted only later when I visited the agricultural division of the University of California at Davis.

This division is at the forefront of agricultural science in the U.S. and is well-known for its leadership role in the field.

From what I was told, the students seem to be the ones leading the faculty in new directions rather than the other way around. When I learned that one of the many projects underway there was a farm run entirely by students and aimed at natural farming, I decided to meet with the students involved in the project. What I said essentially was this:

"I admit that seeing different kinds of primitive crops growing in the middle of symbiotic crops and meadow grasses is interesting, but not enough is being done to make use of key crops such as clover and alfalfa. Growing crops seems to be taking precedence over efforts to revive and invigorate the soil. Frankly, it looks to me as if you're just groping around in the space that separates organic farming and natural farming. At this rate, I doubt that what you're doing will have much impact on the world."

When they heard me say this, the students, led by a young Ethiopian who was their leader, bombarded me with questions. I forget everything that we talked about, but I do remember some of the things the students asked:

"There's something empty and meaningless about agriculture classes. How can one have a rapport with nature through books?"

"Isn't it true that the closer one approaches to nature with natural farming, the less this works to the benefit of mankind?"

"How does one begin to find union with nature? Through observing nature?"

"What is the difference between nature and non-intervention?"

"Do you think that growing primitive crops is the quickest road to nature?"

"You tell us not to apply human knowledge or actions, but doesn't it go against the spirit of natural farming to grow genetically improved eggplants?"

"You talk of speaking with the crops, but do you mean by this that the eggplant is unhappy?"

These and many other odd topics arose. With the difficult questions posed and the unusual answers given, the level of everyone's voices began to rise in the general excitement and before I knew it a large number of onlookers had gathered around the students. I don't recall exactly how I responded to each question asked, but for some reason I have a knack at times like this for immediately giving out offbeat answers. All I do remember for sure is that we had a very enjoyable discussion that day.

I believe I said something to the following effect.

"You listen to your teachers speak about nature in lecture halls under fluorescent lighting. That's why you find the lecture dull. When a man and a woman sit together under the cool shade of a tree on a sunny day and talk, that alone is enjoyable, isn't it? If you want to know whether the eggplant is happy or sad, ask the eggplant. But you ought to be worrying about yourselves rather than the eggplant.

"If people try to grow crops with their minds, they end up as mere farmers. But if they look on with detachment, they are able to observe the entire universe through the crops. The view of the universe and the religious view are one and the same, as are also the view of society and the view of life; they must not be distinct. If you understand the heart of one *daikon*, you understand all. You see that religion and philosophy and science are all one, and that there is nothing.

"People have the attitude that 'I am a priest and so I understand the heart of God but not the heart of the eggplant,' or 'I earn the bread on my table by being a professor of Western philosophy so I have no desire to become a farmer and grow food.' All that is nonsense. The question is from where does the confusion arise in the hearts of those people who say it is not necessary to know the heart of the eggplant or who worry that the eggplant is sad.

"Man has lost sight of the spirit of God and can no longer

know things because he has forgotten that our food, clothing, and shelter are all the creation of God. Man has ventured out and begun wandering in search of knowledge without understanding what it is to know. Human knowledge prevents one from knowing the essence of things; it serves only to cloud and cause us to lose sight of the spirit of things. Because we really have no idea what natural water is, we believe that the water that has passed through the plumbing is the same as river water.

"People talk of creating soil and making water, and then they waste water and use it carelessly, taking it from its true course and utilizing it in ways that were never intended. Nonintervention, or abandonment, could well be called a state in which nature has strayed off course.

"When the spirit of human 'fabrication' is released, when human knowledge is abandoned, nature begins to return to its own true form.

"The revival of nature means more than just returning to the beginning; it means nature creating a new nature. The ultimate aim of my natural farming that returns to nature is the freeing of the human spirit."

When I concluded my talk by saying that this, in simple, easy-to-understand terms was I meant, one young woman with a frown on her face said, "It's so easy to understand that I don't understand it at all." Everybody burst out laughing.

I then said, "It's very simple. Because everything is useless, I'm saying 'just forget what I told you.'" One student made light of this by adding, "What you just said was useless." This evoked another round of laughter. I've never had such a pleasant discussion. When I told the students at the end not to let the professors know what we had talked about, they started chuckling. Two gentlemen walked forward from the circle of listeners with wry grins on their faces. When I shook hands with them and looked at the business cards they had given me, I found that I had just met the president of the college and the head of the agriculture department. They spoke in turn about how my talk had made a deep impression on them and of their concern over the future of the students.

One woman farmer listened carefully that entire day without saying a word. The next day, when I went to have a look at the Sunday market in town, I discovered that she was the organizer of the market. Led by the Ethiopian student, students and others associated with the school were busily selling farm produce at various stands. Many people at the market knew me directly or indirectly and we shook hands.

Three Don Quixotes

Let me describe briefly what happened at the international conferences I attended. The first conference took place at Olympic College in Washington. This campus was a quiet place thickly forested with large trees. The main hall where the conference was held had a very gradually rising steplike structure. I would say that there were about 600 participants.

One thing that took me by surprise was that the remarks made at the opening ceremony were given by a Native American professor at the college. He wore a feather headdress and a beautiful and dignified ceremonial Indian costume. In his talk, he asked what human knowledge is while making reference to old American Indian legends.

Although part of the first day was devoted to introductions of conference attendees from various countries, the main presentations were given by Bill Morrison, who advocates what he calls permanent agriculture in Australia, Wes Jackson of the University of California, who is widely known for the energy-saving agricultural methods that he promotes, and myself.

In contrast with modern agriculture that exploits the soil, Bill's "permaculture" allows farming to be continued permanently on the same land; his method consists of creating farms designed in the form of gardens using perennials and other plants. He seems to be raising quite a following among organic farming advocates throughout Australia and the U.S.

The main thrust of Wes Jackson's talk was that, unless farming methods which reduce the consumption of oil energy to an absolute minimum are widely adopted, farming has no future. He seemed to be searching for new methods of farming while remaining basically supportive of scientific farming.

I was introduced by the host as an advocate of natural farming, which rejects the tenets of modern science and is founded on the philosophy of "nothingness." He added, "It will be interesting to see how Mr. Fukuoka responds to Bill Morrison's method of organic farming and Wes Jackson's scientific farming."

The next day, a panel discussion between the three of us was held, the stated purpose being to find a future course for farming from three people with three differing points of view. We were lined up on the stage and a tournament-like question-and-answer format adopted. A list of the main questions had been given to us the day before, so the debate generally proceeded smoothly, but the affair descended unexpectedly into the absurdly slapstick and brought the audience's laughter down upon us.

Jackson teased Morrison, saying that his accent was awful and that he couldn't understand a word of what Morrison was saying. As for me, I had three people interpreting for me. Every time I said something, this was interpreted three different ways. People in the audience poked fun at this, asking which of the three versions was what I had really said. This showed Americans just how difficult it is to understand Oriental languages and the Oriental mind. The listeners, both perplexed and impressed, posed odd questions, which were followed by just as strange answers amid much laughter all around.

At the end of our debate, I drew a picture of Don Quixote's donkey. Riding on the back of the donkey was

Morrison, who I depicted as blind. A deaf Jackson was riding backwards, while I was hanging onto the donkey's tail and swinging back and forth. I asked the audience: "Here we have three Don Quixotes trying to stop that donkey from running wildly over the cliff by returning to nature, but their efforts are all in vain. What would you do?" Then I drew President Reagan standing on the donkey's neck and dangling a carrot in front of its nose. "What is this carrot?" I asked. One person replied, "Money."

The host laughed at my sketch and said that it would be a shame if this were to sum up the results of the day's meeting. He called the conference to a close and announced that the next conference would take place in San Francisco.

The three Don Quixotes.

The second international conference was held a week later at the agricultural department of the University of California at Santa Cruz. This university had an original layout; it was built on a very large site and the first thing that greets you when you pass through the gates is a broad meadow that stretches as far as one can see. I did not have the impression that I was entering a university campus. This land apparently was once a ranch. The university buildings are located in a large virgin forest of hemlock and redwoods more than 200 feet high and measuring several yards around. Because the forest has been preserved in its original state, the buildings are hidden among these enormous trees and are often hard to locate. It is usually too far to walk to the next building. During the three days of the conference, seminars with participants from many different countries were held in class-rooms in about ten different buildings. Those wishing to attend a seminar were able to board the campus buses that passed by every ten minutes or so. The buses even occasion-ally helped out people who got lost in the woods. Somehow, this sort of system seemed very American to me.

No other seminars were held at the time of my presentation. I was given plenty of time for my talk, on top of which I was supposed to wrap up the conference. The hall was packed with 800 people, half of them backers of the "nature move-ment" gathered from many different countries, and the other half local people connected with the school. I was told to expect both many supporters and opponents of scientific agriculture in the audience. That should give an idea of just how tense the atmosphere was.

I began by talking about what had led me as a young man to become a proponent of natural farming. Wishing to make the philosophical and religious views of my way of natural farming as easy to understand as possible, I explained these using my philosophical cartoons. The cartoons that I drew on sheets of transparent paper with a felt-tip marker were pro-jected onto a large screen behind me for the audience to see.

I also used lots of slides and other visual aids to explain the methods of natural farming and where this now stands. Finally, at the end of my talk, because the conference host had asked me earlier to prepare a closing message for the conference, I had a statement summarizing my conclusions and expressing my hopes translated into English and read for the audience.

I forgot what exactly I spoke about that day, but I do remember exchanging some jokes with the audience; they seem to have enjoyed the talk. However, when the interpreter began reading out my statement, the entire hall suddenly grew hushed. I worried a bit about what would happen. After the statement had been read, the host asked if anyone had any comments and called on two people who raised their hands to give their thoughts.

The first was a man from India. He said that my ideas were exactly like those of Gandhi and, adding that ancient Indian texts mentioned that no-till farming existed there long ago, expressed his support for my way of farming.

The next person to rise was a very influential professor of religion and philosophy at the University of California. The day before, we had come to a disagreement on the question of Western philosophy. He maintained that Socrates was demented. I said that although I could reject everything about the Western philosophers from Descartes on down, I was unable to find the slightest thing wrong with anything that Socrates had said. We parted with the understanding that we would come to some sort of resolution the following day at the meeting. Naturally, I expected him to present an opposing view. He stood up, turned his back to the podium and, facing the audience, launched into a five-minute speech.

"In Western philosophy, Descartes, Locke, Kant, Hegel, and others have explained the process by which the foundations of modern science have been built up. Fukuoka rejects the tenets of Western philosophy and has even succeeded in demonstrating the validity of his stance. This is an astounding feat and modern scientific farming has no choice but to admit that its foundations have been overthrown. I welcome Fukuoka's natural philosophy and farming as the practical and theoretical foundation for a new age."

The end of his speech was met by a thunderous ovation that continued for a long time. This struck a nerve in me too, and I felt at last that I had been right to come to America a second time. A load had been lifted from my shoulders.

Here is the statement that I presented at the conclusion of my talk.

A Statement

(1) *Let Us Return to a Nature Ruled Over by God*
At long last it has become clear that the growth of materialistic civilization does not bring happiness to man. We see now that it both destroys nature and blights the heart of man. Materialistic civilization has expanded to its outer limits based on the thinking that underlies dialectical development. Today we are seeing a period of disintegration and destruction.

The next age must reverse and become an era of spiritual culture that returns inward to a nature ruled over by God. This must be an age of consolidation in which, taking the road of non-action and non-knowing, we elucidate the true nature of man.

We must now bring to an end outward dialectical development and move on to an age of Buddhist contraction and convergence.

(2) *Let Us Halt Science's Wild Rampage*
The scientist must know his own domain and himself take responsibility for halting the wild rampage of science.

Scientific truth can never be absolute truth. Viewed from afar—from the seat of God, scientific truth is always incomplete, consisting only of false conclusions. Changing with space and time, it leads the world astray and is able only to create false goods and give people false joy.

Man must humbly reflect upon the fact that he cannot know and make use of true nature through the accumulation of learning based on discriminating knowledge and through the analysis of nature. What is regarded as "high technology" is really only peripheral technology. Life scientists in particular must awaken from this foolishness that has them toying with the more carapace of natural life and causes them to gallop after euphoric illusions that are nothing but mere shadows of life. The liberties taken with the life sciences are leading to the robotization and debasement of mankind, and pushing humanity closer to its day of reckoning.

(3) We Must Hurry to Establish a Human Charter for the Twenty-First Century

We must today convert religion to religion that transcends religion, return philosophy to its original purity, and replace science with a science that rejects modern science.

To accomplish this, we should hold a world conference bringing together sages from all the religions, philosophies, and sciences, as well as politics, economics, art, and other fields of endeavor, and hurry to draw up a human charter which will serve as a compass to chart the future course that humanity should take.

Unlike other conferences, the purpose of this gathering would not be to argue and debate, creating instead discord and confusion. By bringing together wise men and sages, the purpose would be to show what is truly necessary and what is not; to winnow out and discard learning, science, and heretical religion not needed by man; and to clearly point out the principles that remain as the necessary guideposts for the creation of true men—simple, forthright people of non-action who return to nature.

(4)   *We Must Stop the Desertification of the Earth*
Today, the global loss of vegetation is not something
that can be mended with passive ecological movements to save nature. We must recognize that desertification basically arose from the errors of materialistic civilization and modern scientific agriculture, and
must strive to effect sweeping and revolutionary
changes in agriculture.

Peace and freedom on earth shall arise when nature
and man become one and the earth is transformed
into a paradise filled with vegetation and food. The
grasses and trees have no national boundaries.

Let us scatter the seeds of various drought-hardy
grasses, trees, and food crops by airplane all at once
over those areas on earth that have turned to desert.
This is the only way that the earth can be revegetated.
Let us begin by removing national boundaries with
the trees and grasses. The liberation of peoples everywhere shall begin from this point. That is what I
propose.

Reclaiming the Desert at the Zen Center ──────────

I revisited the Zen center at Green Gulch on the outskirts of
San Francisco that I had been to seven years ago. The area in
which this center is located consists of bare-topped mountains
covered with vegetation reminiscent of the savanna, but a
look at the national redwood forest park located nearby
strongly suggests that this entire area was once densely
forested.

The leader of the Zen center, an important Indian chief who
had shown me around on my first visit to Green Gulch was
no longer here.

The redwood forest covers a wide area with towering trees
200–250 feet high. Seven years ago, when I told him that the
mix of trees growing in the park and the ecology of the under-

brush very closely resembled that of virgin forests in Japan, and that this forest held clues to the revegetation of California, he said, "You may be small, but you are a giant of the Orient."

Explaining that although the redwood grows fast, it has shallow roots and topples easily, I promised to send over the seeds of certain cryptomerias that have deep roots. When I mailed him a packet of seeds after my return to Japan, he sent back as a gift a cup hand-carved from the top of a redwood tree.

I learned by word of mouth that he had carefully planted the cryptomeria seeds. On my second visit to the Zen center last year, the first thing that I was shown was a large photograph of the late director shortly before his death. Surrounded by his followers, he was sitting up in bed and planting seeds in a nursery box. Before dying, he had instructed his followers: "Plant the seeds carefully; they are Fukuoka's soul. When they germinate, plant the young saplings in the valley over there."

When I realized just how dearly he, who with that large body of his reminded me of the recumbent images of the Buddha, had thought of those seeds which I had sent him and how much he too had sought for a way to revegetate the desert, I felt a great lump rise in my throat and was unable to say anything.

I went with more than a dozen people to the place that he had told his followers to plant the saplings. There I found several hundred cedar saplings that had grown to a height of 6 feet or so. Steel posts had been driven into the ground around each one of the saplings and barbed wire strung on the posts about the saplings. I was told that this was to keep the deer away. I saw now that he and his followers had gone to a lot of trouble that I had never even suspected.

"The Great Teacher must surely be delighted because you have come, Mr. Fukuoka," someone said. "He sleeps now on that mountainside over there." The place pointed out to me was a slope on the other side of the valley. I could make out a cairn of small stones perhaps 3–4 square yards in size. It reminded me of the crude graves I had seen in the Somalian desert.

"Mr. Fukuoka, the Great Teacher must be calling out to you: 'Let's sow seeds in the desert.' "

Half in jest, I said, "Looks like a comfortable place to sleep. I wouldn't mind joining him." All of a sudden, I burst into tears. I could say no more.

Yes indeed, here lay someone who was a true sower of seed. When I thought that he may have been the only one who really understood, the only one who would have lived and died together with me, I stood there for a long time not thinking even to wipe the tears streaming down my cheeks.

Why should I, who had not cried even when my own parents died, be crying now? I did not know the reason. I have cried only twice these 50 years. The first time was in 1979 when I gave a lecture at the summer camp at French Meadow. I had begun reminiscing and talking of my conversion that spring day long ago when I was a young man of twenty-five. As I started to ask, "What is true nature?" I suddenly choked on my words and tears welled up. I had to have the talk stopped. The situation now was totally different, but somehow it seemed to me as if these were the same tears.

He is no longer here. Neither his body nor his soul remains in this world any longer. There is no "other world." Knowing that not even that other world in which his soul can wander exists is why I am able to cry.

I sensed that those tears were shed from a point beyond life and death; that I had been bathed, rather, with sweet tears of ecstasy.

The people of the Zen center must have had similar memories. They left me standing alone and, gazing up at the blue Californian sky, slowly made their way back to the Zen center talking happily of him.